THE FAITH OF A PHYSICIST

THE FAITH
OF A
PHYSICIST

*Reflections of a
Bottom-Up Thinker*

THE GIFFORD LECTURES
FOR 1993–4

—

John Polkinghorne

Princeton University Press

Princeton, New Jersey

Published in the United States of America
by Princeton University Press, 41 William Street,
Princeton, New Jersey 08540

Published in Great Britain by the
Society for Promoting Christian Knowledge,
Holy Trinity Church,
Marylebone Road,
London NW1 4DU

Library of Congress Cataloging-in-Publication Data

Polkinghorne, J. C., 1930–
    The faith of a physicist: reflections of a bottom-up thinker:
the Gifford lectures for 1993–4 / John Polkinghorne.
        p. cm.
    Includes bibliographical references and index.
    ISBN 0-691-03620-9 (alk. paper)
    1. Nicene Creed. 2. Natural theology. 3. Faith and reason—Christianity.
I. Title. II. Title: Gifford lectures.
BT999.P65 1994
238'.142—dc20   93-41071

Printed in Great Britain

1 3 5 7 9 10 8 6 4 2

To Ruth

There is an intellectual desire, an *eros* of the mind. Without it there would arise no questioning, no inquiry, no wonder.

Bernard Lonergan

There can be . . . no thought about any reality about which we can do nothing but think.

Austin Farrer

True respect for the mystery [of the incarnation] can express itself, among other ways, just in the attempt to understand it fully.

Wolfhart Pannenberg

For it is the God who said, 'Let light shine out of darkness,' who has shone in our hearts to give the light of the knowledge of the glory of God in the face of Christ.

2 Cor. 4.6

Contents

ACKNOWLEDGEMENTS

It was a great honour for me to be invited by the Principal and Vice-Chancellor, Sir David Smith, and the Senatus Academicus of the University of Edinburgh to give the Gifford Lectures in that University in the academic year 1993–4. I am most grateful for the privilege, and for the hospitality I have received. It was a special pleasure to return in this manner to the University in which I first held a lectureship.

I am most grateful to my secretary, Mrs Josephine Brown, for her unfailing care and skill in typing several drafts of these lectures. I also thank the editorial staff of SPCK for their help in preparing the manuscript for press, and David Mackinder for his patient work on the copy. My wife, Ruth, assisted with reading the proofs, but my causes for gratitude to her far exceed that valuable help, and in grateful recognition of our thirty-eight years of married life together I dedicate this book to her.

John Polkinghorne

The President's Lodge
Queens' College, Cambridge
November 1993

Introduction

It has become fashionable to write books with titles such as *Religion in an Age of Science* (Barbour), *Theology for a Scientific Age* (Peacocke), or *Theology in the Age of Scientific Reasoning* (Murphy). They signify the recognition that the interaction between science and religious reflection is not limited to those topics (such as cosmic history) concerning which the two disciplines offer complementary insights. It involves also an engagement with habits of thought which are natural in a culture greatly influenced by the success of science. To take this stance is not to submit to slavery to the spirit of the age, but simply to acknowledge that we view things from where we stand, with all the opportunities and limitations inherent in that particular perspective. The invitation to give the Gifford Lectures, whose subject is to be 'The Knowledge of God' treated by those who aspire to be 'sincere lovers of and earnest inquirers after truth', provided me with the opportunity to offer my own contribution to this genre. I differ from my predecessors in wanting to make much more detailed contact with the core of Christian belief. My concern is to explore to what extent we can use the search for motivated understanding, so congenial to the scientific mind, as a route to being able to make the substance of Christian orthodoxy our own. Of course, there are some revisions called for in the process, but I do not find that a trinitarian and incarnational theology needs to be abandoned in favour of a toned-down theology of a Cosmic Mind and an inspired teacher, alleged to be more accessible to the modern mind. A scientist expects a fundamental theory to be tough, surprising and exciting.

The chapters that follow present the grounds on which I reach my conclusions. First, I present such understandings of the nature of humanity and of how we may gain knowledge, as result from a critical encounter with the findings and methods of modern science. Then I consider how to speak of God and of creation in the light of contemporary understanding. A coherent concept of divine providential agency is available to us, but its embracing will demand that we take seriously divine engagement with the reality of time in a way which requires some revision of the ideas of classical theology. Both developments arise from the recognition that the physical world is a universe endowed with true becoming.

The central three chapters are concerned with an evaluation of the life, death and claimed resurrection of Jesus Christ. Scientific generality does not impinge directly upon the questions of historical particularity involved, but the approach followed is the one, natural to a scientist, of the search for motivated belief, together with the recognition that the way things actually are often proves contrary to prior expectation and thus enlarges our intellectual imagination. My conclusions are that a belief in the raising of Jesus that first Easter Day, and an adherence to a modified form of kenotic Christology as a means of affirming the meeting of the divine and the human in him, are tenable beliefs in a scientific age. It is the search for an understanding of the work of Christ, evaluated in the light of Christian experience, which controls my Christological account of his nature.

The chapter on the Spirit and the Church also contains some simple thoughts on the reasons for trinitarian belief. While many have apparently felt that some form of evolutionary optimism was a fitting religion for an age of science, I draw a darker conclusion from physical cosmology alone. Only a great new act of God can deliver the universe and ourselves from eventual futility. I discuss and defend the coherence of such an eschatological hope.

An inescapable part of contemporary religious consciousness is the awareness of the stability and diversity of the world's historic religious traditions. My last chapter seeks to explore some of the perplexities which arise from the unavoidable discussion of the urgent theological problem of how we are to understand the way the world faiths relate to each other. I believe that a consideration of how their differing accounts of the status of the physical world relate to the discoveries of modern science, might make a modest contribution to the necessary dialogue.

Before setting to work, it is necessary for me to defend the propriety of regarding the exercise as one which falls within the provisions of Lord Gifford's will.

It may seem to be rather strange, or even to smack somewhat of theological sleight of hand, that the creed of a scientist turns out to be the Nicene Creed. The discussion in all but the final chapter of these Gifford Lectures is woven around phrases chosen from that creed.[1] My defence of this must rest, of course, on the detailed arguments that follow, but it is perhaps not altogether surprising that the distillation of four centuries of Christian

[1] I do not consider every phrase of the creed. The omission of 'one baptism for the remission of sins' does not mean that I consider this unimportant, only that I do not feel I have anything particularly new to say about it.

experience and intense intellectual debate should prove to have lasting value. The creed is very spare in its formulation. It provides a framework for Christian thought. The way I fill in that outline is influenced, both in style and in content, by my experiences as a scientist (more particularly, a theoretical elementary particle physicist) and as a twentieth-century Christian believer who has spent all his life within the contemporary worshipping and confessing community of faith. Let me offer some explanation of why I believe this procedure of credal exploration to be a fulfilment of the will of Lord Gifford.

He required his lecturers to engage in 'Promoting, Advancing, Teaching and Diffusing the study of Natural Theology' and he bade them to 'treat their subject as a strictly natural science, the greatest of all possible sciences . . . I wish it to be considered just as astronomy or chemistry is'. While the lecturers are told to be 'under no restraint whatever in their treatment of this theme', they are also enjoined to frame their discourse 'without reference to or reliance upon any special exceptional or so-called miraculous revelation'.

Judgement that I have indeed complied with these requirements must depend upon certain understandings of the nature of natural theology and the nature of revelation. I have addressed these issues in earlier writings,[2] so I can be brief on this occasion. I see natural theology as an integral part of the whole theological quest for understanding and by no means an isolable or merely preliminary sub-department of it. Such a view can fittingly be maintained in Edinburgh, where it has long been defended with great vigour and learning by Professor Thomas Torrance.[3] Natural theology derives from the general exercise of reason and the inspection of the world. It is part of theology's necessary engagement with the way things actually are. While it will have its own areas of focused concern (such as aspects of the doctrine of creation), it will also impinge upon other matters less traditionally associated with it. For example, such concepts as we may form of the nature of humanity and its relationship to God, will exercise an influence upon how one approaches and understands the Christian doctrine of the incarnation of Christ. I have subtitled this series 'Theological Reflections of a Bottom-Up Thinker', for my intention, as I have explained, is to explore, as far as I am able, how one who takes modern science seriously, and whose habits of thought – for good or ill – are formed by long experience of working as a theoretical physicist, approaches questions of the justification and under-standing of religious belief. For me, doing natural theology is part and parcel of doing theology. The fundamental question to be asked about any

[2] Polkinghorne (198°, ch. 1; (1991), ch. 4.
[3] Torrance (1969); (1985).

theological statement is, 'What is the evidence that makes you think this might be true?' Of course, the kind of evidence considered, and the kind of understanding attained, must be conformed to the Reality about which one is attempting to speak, but so it is in the natural sciences also. Ethology and elementary particle physics use different techniques and employ different categories precisely because their subject material is so different. 'Exploration into God' will call for its own procedures and insights, but that should not lead us to deny to the intellectual reflection upon it the honourable title of 'theological science' – in Lord Gifford's view (and mine) 'the greatest of all possible sciences'.

David Jenkins wrote of theology that its

> understanding is not and never can be a theory which provides a frame which embraces everything. Rather it is, and must always be, open to new and hitherto unknown possibilities. In this I am convinced that true Christianity is precisely the same as true science. Both are required to be totally open to whatever is authentically given in each situation. This is no mere coincidence. For it is Jesus Christ who definitively makes it clear that the universe is truly open to truly scientific investigation.[4]

This last remarkable assertion comes from a book whose basis is to explore the consequences of taking seriously the fact that our universe has evolved persons, so that 'we are confronted with a coincidence of fact and value which shows every sign of being an identity and which requires an investigation and responses proper to the reality which is perceived'.[5] Jenkins believes that only God's existence and love can deliver humankind from absurdity and so render the universe truly an intelligible cosmos, and he says of Jesus that '*This* data and no other is finally decisive for our true understanding of man and the world.'[6] This is a modern version of what in the second century would have been called the Logos-doctrine. These are issues I too wish to pursue in these lectures, in my own way.

It is by seeking to start with the phenomena that give rise to the theories, that I characterize myself as a bottom-up thinker. It is a natural stance for a scientist to adopt. We have learned so often in our exploration of the physical world that 'evident general principles' are often neither so evident nor so general as one might at first have supposed. Many theologians are instinctively top-down thinkers. I do not deny a role for such ambitious intellectual effort. I am merely wary of it and wish to temper its grand generality with the questionings that arise from the consideration of

---

[4] Jenkins (1967), p. 78.
[5] ibid., p. 8.
[6] ibid., p. 41.

particularity. Perhaps chapter 9 illustrates most clearly the distinction I have in mind. Of course, as a believer, I wish to embrace the great statements of Christian eschatological hope, but I also need, however tentatively and speculatively, to ask how such hopes relate consistently to what we know of the process and history of this present world.

I am conscious of how much modern theological thinking appears narrowly parochial, concentrating solely on humanity and taking a view of history which effectively covers the span of a few thousand years rather than the billions of years of the universe's past and future evolution. I do not deny the singular significance of the emergence of men and women from the womb of cosmic history, and it is a principal purpose of these lectures to explore the meaning of Jesus of Nazareth as the clue to divine purpose in that history. Yet, when Stanley Grenz tells us of a distinguished contemporary theologian that 'Pannenberg implicitly makes the seemingly audacious claim . . . that human history is the history of the cosmos, for the consummation of anthropology in the kingdom of God is at the same time the consummation of creation in its eternal praise to the Creator',[7] I am uneasy that our vision is becoming unduly confined and theological discourse is in danger of becoming a little too cosy. The whole of this vast universe must matter to God, each part in its appropriate way, and the human story is but a part of that.

Of course, science has its self-chosen parochialities too. Its great success is purchased through the modesty of its ambition, restricting the phenomena it is prepared to discuss to those of an impersonal, and largely repeatable, character. It was a brilliant tactic of investigation for Galileo and his successors to confine themselves to the primary quantitative questions of matter and motion, but that narrow view would be a poor metaphysical strategy, condemning one to a narrow reductionist conception of reality. Those discarded secondary qualities of human perception may in fact prove to be primary clues to the construction of an ampler view of the way the world is. Music is more than vibrations in the air.

It seems to me that many educated people in the Western world view religious belief with a certain wistful wariness. They would like some sort of faith, but feel that it is only to be had on terms which amount to intellectual suicide. They can neither accept the idea of God nor quite leave it alone. I want to try to show that although faith goes beyond what is logically demonstrable – and what worthwhile view of reality does not? – yet it is capable of rational motivation. Christians do not have to close their minds, nor are they faced with the dilemma of having to choose between ancient faith and modern knowledge. They can hold both together. Revelation is

[7] Grenz (1990), p. 206.

not the presentation of unchallengeable dogmas for reception by the unquestioning faithful. Rather, it is the record of those transparent events or persons in which the divine will and presence have been most clearly discernible. Let us pursue Lord Gifford's chosen analogies with chemistry and astronomy. The laws of chemistry are always operative, but their nature may most clearly be perceived in those well-chosen and contrived events we call experiments. God is always present and active in the world, but it may well be that he is most clearly to be seen in the particularities of what the Judaeo-Christian tradition calls salvation-history. That history is exceptional in the clarity with which the divine can be recognized through it, not in an implied absence of God from other times and places. The need to seek God where he can most clearly be seen has the consequence that the unique is not to be excluded from our consideration. Just as the astronomer finds an unrepeatable regime of the greatest significance in the reconstruction of those highly energetic moments following the big bang, so the Christian theologian finds a regime of unique significance in the reconstruction of the life and death and resurrection of Jesus Christ. For me, the Nicene Creed is not a demand for intellectual surrender to a set of non-negotiable propositions; instead it represents the summary of insights and experience garnered from the founding centuries of the Church's history. I therefore feel that its critical exploration is a fitting way of fulfilling Lord Gifford's intentions. To eschew the use of this material would be to act as foolishly as did those savants who declined to look through Galileo's telescope. Opportunities for gaining insight are not wilfully to be refused.

Yet the task is formidable, and I cannot claim completeness in its execution. What I can aspire to is a candid and honest attempt to explore the foundations of Christian belief and to try to offer explanation for that belief comparable to the kind of explanation one might offer of one's conviction that matter is composed of quarks and gluons and electrons. In both cases, a web of interlocking insights has to be woven before the tapestry of understanding can be presented for inspection. The Nicene Creed is the loom on which I seek to weave my tapestry. If the warp is the engagement with the record of Christian tradition, the woof is the engagement with contemporary understanding of ourselves and the universe we inhabit. Neither should command a slavish and uncritical assent; both will leave important issues short of full resolution. In both science and theology we encounter questions where we have to confess that we do not know the answers. Many of these questions relate to the mystery of human personality. We should not be surprised to find the divine nature even further beyond our grasp. To acknowledge these limitations is not to abandon the task, simply to be modestly realistic about what we can achieve.

Those who labour on the border of science and theology employ different styles in their work. Some produce magisterial surveys of different points of view, laying out the options before the reader. The *doyen* of such writers is Ian Barbour. He is a kind of modern Master of the Sentences, as his Gifford Lectures showed.[8] I personally prefer a more selective approach, focused by the Nicene lens on central Christian issues. As best I can, I want to build up that synthesis which seems to me to make the most sense and to hold the most promise. Of course I profit greatly from the work of others, and make quite extensive quotations to avail myself of their insights, and of course I try to take account of the criticism of those whose views are opposed to the one I espouse, yet my aim is not to survey the scene but to propose an interpretation. I am convinced that the discussion must not just be on the frontier between science and theology, but must penetrate as deeply as possible into their heartlands. I am not really interested in a lowest-common-denominator 'rational religion'. What I want to know is whether the strange and exciting claims of orthodox Christianity are tenable in a scientific age. Hence my engagement with the Nicene Creed in an attempt to address the issue.

Science and theology relate to their past in different ways. Because of its ability to manipulate and interrogate its material, science conquers territory over which it gains perpetual sovereignty. A well-explored physical regime can be mapped in a way which will not call for subsequent revision. To a certain degree of accuracy and level of detail, Newton has said all that needs to be said about gravity and the solar system. His methods are sufficient for sending a space-probe to Mars. Yet, in another sense, he did not utter the last word, for even his achievements could amount to no more than verisimilitude. Investigation of phenomena on a finer scale of accuracy led to the confirmation of Einstein's theory of general relativity. The latter subsumed Newtonian physics rather than abolishing it, for it was essential to annex the latter's successes, explained as limiting cases in an appropriate approximation. This nesting relationship of successive scientific theories gives the subject its character of a cumulative advance of knowledge. A very ordinary scientist today possesses, in consequence, much greater overall understanding of the physical world than was ever possible for Sir Isaac.

The situation in theology is entirely different. The Object of its study is not open to manipulation, nor can he be caught in our rational nets. Every encounter with divine reality has the character of gracious gift and it partakes of the uniqueness inherent in any personal meeting. The theologian of the twentieth century enjoys no presumptive superiority over the theologians of the fourth or sixteenth centuries. Indeed, those earlier centuries may well have had access to spiritual experiences and insights

---

[8] Barbour (1990).

which have been attenuated, or even lost, in our own time. I do not wish to deny that our age offers its own particular opportunities and intellectual understandings (not least, in the conditions of consonance which our enhanced understanding of the history and structure of the physical world impose upon theological discourse about creation), or to suggest that we can be anything but twentieth-century thinkers. Yet we must maintain a dialogue with the past as a corrective to the limitations of the present. That is why I want to explore in these lectures the extent to which traditional Christian belief can be seen to be valid when it is critically examined in a scientific age. If it could not stand up to that scrutiny, that would be that, but in fact I shall be suggesting in what follows that the Nicene Creed provides us with the outline of a rationally defensible theology which can be embraced with integrity as much today as when it was first formulated in the fourth century.

Throughout, my aim will be to seek an understanding based on a careful assessment of phenomena as the guide to reality. Just as I cannot regard science as merely an instrumentally successful manner of speaking which serves to get things done, so I cannot regard theology as merely concerned with a collection of stories which motivate an attitude to life. It must have its anchorage in the way things actually are, and the way they happen. I can agree with Keith Ward when he says, 'We need to have a picture, a focal concept, of the nature of divine love as sharing and self-giving'; but I cannot go on to say with him, 'but that picture could be given in an inspired story, as it is with Rama and Krishna in the Indian tradition'.[9] A bottom-up thinker is bound to ask, What makes you think this story is a verisimilitudinous account of Reality? The anchorage of Christianity in history is to be welcomed, despite its hazards. For me, the Bible is neither an inerrant account of propositional truth nor a compendium of timeless symbols, but a historically conditioned account of certain significant encounters and experiences. Read in that way, I believe it can provide the basis for a Christian belief which is certainly revised in the light of our twentieth-century insights but which is recognizably contained within an envelope of understanding in continuity with the developing doctrine of the Church throughout the centuries.

I know that God is neither male nor female, but I have to use a pronoun for the Deity, and I follow tradition and convention in using 'he'.

[9] Ward (1991), p. 92.

1

Humanity

'We believe . . .'

At the outset of the endeavour to articulate the creed of a scientist it is necessary to place one's metaphysical cards on the table. It may be a rather scanty hand, short on aces and court cards, but honesty requires that it be exhibited. Some of the players may deny holding any such cards at all, but anyone who tells you that 'The Cosmos is all that is or ever was or ever will be',[1] in that meaningful tone of voice which implies equating the Cosmos with what the physical cosmologists can tell us about what is going on, is certainly making a metaphysical claim. A biochemist, however distinguished, who tells you that 'Anything can be reduced to simple, obvious mechanical interactions. The cell is a machine. The animal is a machine. Man is a machine'[2] is certainly going beyond (*meta*) his biochemistry. The fact is, of course, that none of us can do without metaphysics. We all need to form a world-view going beyond the particularities of our individual disciplines. Scientists are especially prone to recoil from the notion of what they fear will prove to be the cloudy claims of such a generality, and then go on to promote the insights of their own field of study into a rule for all. As Jeffery Wicken says, 'Although scientists may officially eschew metaphysics, they love it dearly and practice it in popularized books whenever they get the chance.'[3] If we are going to be metaphysicians willy-nilly, let us at least be consciously self-critical about it.

Ian Barbour[4] has identified three metaphysical implications of current physics: (1) temporality and historicity (the physical world is endowed with true becoming); (2) chance and law (the intertwining of regularity and randomness as the basis of fruitfully evolving process); (3) wholeness and emergence (increasing complexity of organization gives rise to wholes which

---

[1] Sagan (1980), p. 4.
[2] J. Monod, quoted in Barbour (1990), p. 6.
[3] J. Wicken, *Zygon* 24, p. 162.
[4] Barbour (1990), pp. 123–4.

cannot adequately be described in terms of their parts alone). These properties have often been discussed.[5] Naturally Barbour recognizes the inadequacy of a metaphysics which simply uses physics as its springboard into the beyond. For theology, the critical metaphysical issues are the nature of humanity and the coherence and plausibility of the concept of God, issues in whose consideration physics will play only a modest role. Their discussion calls for much wider reference and its undertaking is no light task, not least because none of these issues is wholly independent of any other and so they must eventually be taken together. Were the universe merely mechanical, and human beings nothing but machines, then the Cosmic Clockmaker, who is the only conceivable God for such a world, would prove an unsatisfying and ultimately otiose deity, as the eventual sterility of eighteenth-century deism showed only too clearly.

The task in prospect is intellectually daunting, but a scientist should summon up the courage to undertake it. What he or she faces in metaphysics is the ultimate form of the search for a theory of everything, the desire to unite knowledge into a satisfying whole. Physicists should be the last people to settle for the 'metaphysical modesty'[6] which resists pursuit of a total understanding. Their instinct is to go for the Grand Unified Theory.[7] Having said that, it is desirable to be realistic about the degree of success likely to be attainable. Even in physics GUTS are not easy to find.

Despite my assertion of their eventual interdependence, one has initially to consider topics one by one. The focus of this chapter is upon our understanding of human nature. Thomas Nagel begins his own attack upon the mind-body problem by acknowledging that what he has to say may well be 'nothing more than pre-Socratic flailing about'.[8] Anaximenes did not get it right when in the sixth century BC he suggested that all matter was made of air in a variety of denser or more rarefied states. Hindsight suggests both that it was hopelessly ambitious for him to have tried to crack the problem by a stroke of guesswork, and also that it was a brilliant intuition to surmise that behind the variety of the physical world's appearance there might be just a limited kind of basic physical stuff. We elementary particle physicists can salute Anaximenes as one of the first of our long line.

Today it may well be hopelessly ambitious to try to crack the problem of mind-body, which lies at the heart of our metaphysical perplexity about

[5] Bartholomew (1984); Polkinghorne (1986), ch. 6; (1988), ch. 3; Prigogine (1980); Prigogine and Stengers (1984); for analogies in biological science, see Peacocke (1979), chs 3 and 4; (1986), chs 2 and 3.
[6] Gilkey (1969), p. 223.
[7] Davies (1984); (1989), ch. 15.
[8] Nagel (1986), p. 30.

humanity. Yet we have to make shift to flail around as best we can, and there may even be some moderately hopeful directions in which to wave our arms. How are we to choose the line of our attack?

The world of thought divides into top-down thinkers, who place reliance upon general principles and pursue their clear and discriminating evaluation, and bottom-up thinkers, who feel it is safest to start in the basement of particularity and then generalize a little. Plato versus Aristotle, one might say. As a physicist my sympathies are with the latter. (I belong to the generation of theorists who plodded along in the wake of the experimentalists, trying to make sense of their discoveries, in contrast to the contemporary young turks who hope to make a killing at one blow by the high-principled application of superstring theory.) In fact, however, one needs a little of both approaches, neither scorning the aid of the specific nor refusing the boldness of essaying an occasional general speculation. A. N. Whitehead put it well when he wrote: 'The true method of discovery is like the flight of an aeroplane. It starts from the ground of particular observation; it makes a flight in the thin air of imaginative generalization; and it again lands for renewed observation rendered acute by rational interpretation.'[9] I think one might start one's flight into the metaphysical empyrean from the runway marked 'consciousness'.

The existence of consciousness is a fact of fundamental significance about the world in which we live. Each of us experiences it, and only the most sceptical of philosophers would question that we rightly extrapolate from our individual perception of it to the belief that its possession is shared by other humans and, to a lesser degree, by the higher animals. It is one of our glories; Pascal wrote that 'All bodies, the firmament, the stars, the earth and its kingdoms are not worth the least of minds, for it knows them all and itself too, while bodies know nothing.'[10] The essence of consciousness is awareness rather than mere ratiocination. A computer can perform incredibly complex logical operations, but only if we were convinced that it had attained awareness would we feel it ethically unacceptable to pull out the plug connecting it to its power supply. It is a highly disputable question whether a computer could reach a degree of complexity which would generate awareness, and perplexing to conceive what would be the test to assure us that it had done so. The assessment would surely depend more upon intuitive empathy than logical analysis in any Turing-test conversational encounter, for in self-consciousness we are getting close to the centre of the mystery of personhood.

Erwin Schrödinger once said, 'Although life may be the result of an

[9] Whitehead (1978), p. 5.
[10] B. Pascal, Pensées (Penguin Books, 1966), p. 125.

11

accident, I do not think that of consciousness. Consciousness cannot be accounted for in physical terms. For consciousness is absolutely fundamental. It cannot be accounted for in terms of anything else.'[11] This led him to the notion of a universal consciousness in which all participate. I cannot follow that path, for it runs contrary to our experience of individuality; but certainly a world-view which takes no account of consciousness would be woefully inadequate. The lunar landscape of a reductionist science, dismissive of all but the material, is not the home of humanity. Keith Ward says of materialism that it is 'the victory of the abstract over the concrete, of the simple over the complex, of brute fact over intelligible necessity'.[12] It flies in the face of our direct experience of mind and treats as uninteresting what is in fact the most significant development of cosmic history: a universe become aware of itself.

I would be of the same opinion about any account of reality which did not find room for what the medieval scholastics called 'intentional being'. Human freedom is to be understood in the strong sense of the freedom to do some act and the freedom not to do it (the so-called 'liberty of indifference'), and not just in the weaker sense of acting willingly ('liberty of spontaneity'), which is compatible, as Luther saw clearly, with an immutable necessity, since the desire and the action could be part of the same causal nexus.[13] When we face a decision, we face a genuine choice; hence our intuition of moral responsibility. I nail my colours to the metaphysical mast with that blunt assertion.

The debate about human freedom or determinism has a long history, but I cannot myself believe that debate to have been a sequence of mouthings by automata, rather than a rational discussion. The exercise of reason is closely allied to the exercise of freedom, for, to put it crudely but directly, if the brain is a machine, what validates the programme running on it? Doubtless the pressures of evolutionary necessity would ensure (and coding in DNA would transmit) a certain rough-and-ready correspondence of thought and reality, but the subtlety and fruitfulness of human reason seems clearly to call for something much more profound for its explanation than just a spin-off from the struggle for survival. With Thomas Nagel I believe that 'An evolutionary explanation of our theorizing faculty would provide absolutely no confirmation of its capacity to get at the truth.'[14] Such a capacity requires human rational judgement to enjoy an autonomous validity which would be negated if it were the by-product of mere physical necessity. John

[11] Moore (1989), p. 252.
[12] Ward (1982a), p. 93.
[13] Kenny (1979), p. 73.
[14] Nagel (1986), p. 79.

Macquarrie says rightly that 'The defence of freedom is . . . that it is the presupposition of every science, investigation and argument. The denial of freedom is a self-contradiction, the *reductio ad absurdum* of the arguments that lead to the denial.'[15] We should not saw off the rational branch on which we seek to sit.

Thus I take the existence of rationality and free will to be part of the foundation on which to build a bottom-up metaphysics. The strategy is to take with equal seriousness all parts of basic human experience. I can see that critics might object to the label 'basic' as begging the question, but I object in turn that there is an implausibility in those who seek to reduce parts of such experience to the status of the epiphenomenal, an implausibility repeatedly exemplified by our inability outside our studies to live other than as people endowed with free agency and reason. Those who make physical reduction-ist claims are also indulging in what the sharp-tongued physicist Wolfgang Pauli called 'credits for the future', vague hopes that things might one day be shown to be that way. In the absence of such demonstration we do well not to submit our world-view to Procrustean over-simplification. With Thomas Nagel I would, for example, want to 'regard action as a basic mental or more accurately psychophysical category – reducible neither to physical nor to other mental terms'.[16]

The direction our thought is taking is that of an ample and many-valued view of human nature, which resists men and women being confined to consideration under some impoverished rubric such as genetic survival machines or computers made of meat. Our nature cannot be objectified in such reductionist terms. Heidegger said of *Dasein* (present being) that it 'does not have properties but possibilities'.[17] There is an openness to it. I would want to go further, beyond the recognition of free agency and intellect, to discern a component of human life which calls for labelling as spiritual. By that I mean simply that there are aspects of our experience which hint at an incompleteness in what we are and that encourage the expectation of a fulfilment whose ground could only be in something or someone other than ourselves. Peter Berger has drawn our attention to 'signals of transcendence' found in everyday life: (a) an argument from order (essentially the intuition that history is not a tale told by an idiot; the parental role of comforting a frightened child is not the acting of a loving lie); (b) an argument from play (cheerfulness, not to say joy, keeps breaking in); (c) an argument from hope (something is held to lie in the future which is necessary to the completion of the present); (d) an argument from

[15] Macquarrie (1982), p. 17.
[16] Nagel (1986), p. 111.
[17] Quoted in Thiselton (1980), p. 186.

13

damnation (our outrage at Hitler and Stalin is an intuition of the transcendent moral seriousness of the world); (e) an argument from humour (there is a perceived incongruity in our experience which 'reflects the imprisonment of the human spirit in the world').[18] I would want to add to these an argument from mathematics.[19] The nature of that subject is a hotly disputed philosophical question, but for many of its practitioners its pursuit has the character of discovery rather than construction. They would agree with St Augustine that 'men do not criticise it like examiners but rejoice in it like discoverers'.[20] Here is the intimation of an independent world of everlasting truth which we are able to explore.

To these signals of transcendence may be added a less focused recognition of unbounded aspiration in the face of human finitude, a stubborn refusal to give the last word to human insignificance on the cosmic scene. For the believer this may lead to a casting of oneself upon God in submission to that feeling of absolute dependence which Schleiermacher thought was the essence of religion. For atheists like Jacques Monod[21] or Steven Weinberg[22] it may lead to the not ignoble stance of heroic defiance in the face of a universe perceived as hostile or farcical. For many in the Western-educated world today it may lead to a kind of wistful fellow-travelling with religion, able neither to accept it nor wholly to dismiss it, retaining a memory of old tales of deity kept echoing in the caverns of the mind more by poetry than by argument. There is a kind of God-shaped hole in many people's lives, and Langdon Gilkey is right to say that 'One of the most striking things about our human existence in this epoch is that we notice this relation to the unconditional as much by its absence as its presence.'[23]

Such considerations as these persuade me that our account of humanity will have to make room for more than our temporal experience of mind and matter. It will have to accommodate that dimension of openness to something beyond us which I have called spiritual, and which carries in the midst of time the hint of eternity. In the words of Diogenes Allen, it is 'the domain of the heart' that we are entering here. Its exploration 'is related to the human quest for life. The intellect is involved in the quest but what is at stake is our own person in what we are, what we ought to be, what we may become, what we may hope for.'[24] He goes on to say later that 'it is those

[18] Berger (1970), ch. 3.
[19] Polkinghorne (1988), pp. 75–6. See also, Mickens (1990) and cf. Barrow (1992), ch. 6.
[20] Quoted in Kenny (1979), p. 15.
[21] Monod (1972), ch. 9.
[22] Weinberg (1977), ch. 8.
[23] Gilkey (1969), p. 309.
[24] Allen (1989), p. 13.

who are not seekers who must account for not being so since there are fundamental questions concerning the existence and order of the universe that are vitally important to how we shall live and what we shall hope for'.[25]

Yet there is also a dark side to human nature, that inherent flaw by which our better aspirations are frustrated. Introspection and the study of history alike show a slantedness towards corruption and frustration. 'So I find it to be a law that when I want to do right, evil lies close at hand' (Rom. 7.21). A country's liberator so often becomes its next tyrant. In Christian terminology, we are concerned with sin, which Reinhold Niebuhr once described as being the only empirically verifiable Christian doctrine. Partly this baleful influence resides in human structures. Gutiérrez says that 'Sin is evident in oppressive structures, in the exploitation of humans by humans, in the domination and slavery of peoples, races, and social classes.'[26] But we each of us know that it is also present in the individual fearful human heart. The powerful myth of Genesis 3 diagnoses the origin of this malady in an alienation from God. It is suggested that we are not by nature self-sufficient beings, but in our heteronomy we need to find the way of being reunited with the Ground of our being. The fall is not to be understood as a single disastrous ancestral act from which all our troubles flow. Yet in the course of human evolution there must have been a period of dawning consciousness of the self, accompanied by dawning consciousness of God, in which the former was asserted against the claims of the latter. The consequences of that turning away from the divine presence would find embodiment in resulting cultural and social structures, thereby propagating from generation to generation an influence reinforcing the false assertion by the self of its autonomy. It is even conceivable that this would bring about a genetic bias towards a certain kind of human nature. I reject a strong sociobiological account of genetically determined human behaviour, but there are surely some genetic leashes on which we are held. In this way one can understand today what is meant by the traditional theological concept of an entail of human sinfulness from which we need deliverance by God's grace.

The tenor of the discussion so far has been not to deny our immersion in physical process, nor to deny our evolutionary origin from among the animals, but to acknowledge also that in humanity something new has come into being. At the simplest level, as Jacob Bronowski said when commenting on the work of ethologists and behavioural psychologists attempting to assimilate human and animal behaviour to each other, 'But they cannot tell us everything. There must be something unique about man because otherwise, evidently, the ducks would be lecturing about Konrad Lorenz

[25] ibid., p. 214.
[26] Gutiérrez (1988), p. 103.

15

and the rats would be writing papers about B. F. Skinner.'[27] At a more profound level, George Steiner says, 'man alone can construct and parse the grammar of hope . . . More especially: of all evolutionary tools towards survival, it is the ability to use future tenses of the verb – when, how did the psyche acquire this monstrous and liberating power? – which I take to be foremost.'[28] Humanity is open to wide horizons of reality, a fact which finds its reflection theologically in our being sinful, redeemable, worshipping beings.

This character of spiritual openness possessed by humanity stands in some tension with our understanding that we have emerged in the course of a history of continuous development linking the universe today with the expanding and rapidly transmuting fireball consequent upon the big bang. There does not seem anything very spiritual about an energetic quark soup (the universe 10^{-10} sec old) or the amino-acid-rich shallow pools of early Earth. So where did the spiritual come from? Arthur Peacocke tells us that 'we see a world in process that is continuously capable, through its own inherent properties and natural character, of producing new living forms – matter is now seen to be self-organizing',[29] and he goes on to say later that 'we cannot avoid arriving at a view of matter which sees it as manifesting mental, personal and spiritual activities'.[30] I agree with these words, but they are more a statement of the problem than its solution.

Anyone expressing reservations about the total adequacy of the neo-Darwinian account of evolutionary history is apt to be looked at askance, as if he were a crypto-creationist seeking a gap into which to insert a pseudo-deity. Yet one can accept the insights of natural selection and still feel that one has not heard the full story. There are two major problems. One is the question of timescale. Three or four billion years may seem a pretty long time for the coming to be of life and the formation of its evolved complexity, but incredibly intricate developments have to be fitted into that period. Someone like Richard Dawkins[31] can present persuasive pictures of how the sifting and accumulation of small differences can produce large-scale developments, but, instinctively, a physical scientist would like to see an estimate, however rough, of how many small steps take us from a slightly light-sensitive cell to a fully formed insect eye, and of approximately the number of generations required for the necessary mutations to occur. One is only looking for an order of magnitude answer, comparable in crudity to the back-of-the-envelope calculations of early cosmologists, but our biological

[27] Bronowski (1973), p. 412.
[28] Steiner (1989), p. 56.
[29] Peacocke (1986), p. 54.
[30] ibid., p. 123.
[31] Dawkins (1986).

16

friends tell us, without any apparent anxiety, that it just can't be done. So much of evolutionary argument seems to be that 'it's happened and so it must have happened this way'.

The second difficulty is more fundamental. Why do things get more complex with time? Why do multicellular plants and animals emerge when single cellular organisms seem to cope with the environment satisfactorily? There is a direction of increasing complexity apparent in nature. Paul Davies has called this progressive tendency 'the optimistic arrow' of time,[32] in contrast to the pessimistic arrow of the second law of thermodynamics, which entails increasing entropy in closed systems. Our recognition that biological entities are open systems, exporting entropy into the environ- ment, and the insights gained from the study of the thermodynamics of systems far from equilibrium,[33] mean that we understand why there is not a contradiction between these two arrows of time, but that by itself does not explain the existence of the optimistic arrow. The theoretical biologist John Maynard Smith has admitted that 'there is nothing in neo-Darwinism which enables us to predict a long-term increase in complexity'.[34]

Nowhere are these problems more acute than in attempts to understand the rapid and remarkable evolution of the hominid brain.[35] A. R. Wallace, Darwin's contemporary and the co-discoverer of the principle of natural selection, wrote: 'Natural selection could only have endowed savage man with a brain a little superior to an ape, whereas he actually possesses one little inferior to that of a philosopher.'[36] Thomas Nagel takes up the theme when he writes:

> The question is whether not only the physical but the mental capacity to make a stone axe automatically brings with it the capacity to take each of the steps that have led from there to the construction of the hydrogen bomb, or whether an enormous excess of mental capacity, not explainable by natural selection, was responsible for the generation and spread of the sequence of intellectual instruments that has emerged over the last thirty thousand years.[37]

He goes on to say, 'Why not take the development of the human intellect as a probable counterexample to the law that natural selection explains everything, instead of forcing it under the law with improbable speculations

[32] Davies (1987), p. 20. This book provides a useful survey of relevant issues and scientific insights.
[33] Peacocke (1986), Appendix; Prigogine (1980); Prigogine and Stengers (1984).
[34] Quoted in Davies (1987), p. 112.
[35] Eccles (1989).
[36] Quoted in Barbour (1956), p. 92.
[37] Nagel (1986), p. 80.

unsupported by evidence?'[38] Nagel asserts that one does not need to possess an alternative theory before declining in this case to accept explanation by natural selection, but I would want to suggest that the sensible and hopeful alternative direction in which to look is to the existence of higher-order organizing principles, at work in the history of the world. I shall return to this point in due course when we consider creation.

Meanwhile we should acknowledge that an adequate view of humanity will have to take into account the fact of death. Once again we confront the tension between our embedding in the history of the physical world and that dimension of unbounded aspiration which I have called the spiritual. In one sense nothing seems more final than death. Life has gone; the corpse is just the simulacrum of the person we know, and that itself will soon be dissolved into decay – '. . . you are dust, and to dust you shall return' (Gen. 3.19). Yet the human spirit defies death – 'Death be not proud . . .'. Thoughts of transience are the triggers of Intimations of Immortality. (It is no accident that we turn to the poets at this point.) It is not only that we encounter the vertiginous incongruity of our own extinction – 'My death in the world is easy to think about; the end of my world is not'[39] – but we rebel at the thought that the sun's explosion and the eventual collapse or decay of the whole universe will render futile Shakespeare and Mozart and St Francis, and all that they achieved. Once again this is a point to which I shall return.

Hamlet says, 'What a piece of work is a man! How noble in reason, how infinite in faculty', and yet a 'quintessence of dust'. An account of humanity will have to marry these characteristics with metaphysical adequacy if it is to do justice to 'the paragon of animals'. The problem is too complex to admit a reductionist solution, either of a physicalist type, asserting the mental to be at most an epiphenomenon of the material (thought is what the brain does), or of an idealist type, asserting the physical to be a construct of the mind (ideas are the true reality). The former totally fails to bridge the gap yawning between talk of the activity of neural networks, however complex, and the basic mental experience of perceiving a patch of pink,[40] while the latter fails to account for our encounter with the stubborn facticity of the physical

[38] ibid., p. 81.
[39] ibid., p. 225.
[40] The view expressed by Searle (1984, ch. 1) that 'mental phenomena are just features of the brain' does not adequately recognize this distinction. He appeals to analogies with macro- and micro-physics, but here both descriptions are concerned with the same kinds of phenomena, namely those involving energy and momentum. (See the Note on Reductionism at the end of this chapter.) Another point, emphasized by Hodgson (1991, ch. 6), is that if consciousness is nothing more than complex input-output processing of information, what is its evolutionary advantage over the unconscious execution of the same procedures?

world, often so surprising and counterintuitive in its nature and history, and our ability to agree a common description of it. I am not disposed to deny the reality of a long cosmic history during which no minds were present in the universe. Whatever metaphysical resolution we may arrive at, it will have to be even-handed in its treatment of our irreducibly different experiences of the physical and the mental.

Basically there are two alternative strategies that can be followed. One treats the mental and the physical as entirely distinct kinds of reality. This is the approach of dualism, and its modern patron saint is René Descartes. Its strength is its frank recognition of the bipolar character of our experience; its great weakness is its inability to explain the relationship between these two varieties of experience. How does it come about that the mental act of deciding to raise my arm results in the physical act of its actually being raised? Or how does it come about that the intake of drugs can so decisively affect my mental experiences? Cartesians were eventually driven to the desperate remedy of an appeal to occasionalism – God continually brings about a synchronization of the mental and the physical to produce the appearance of a causal connection between their disjoint worlds.

These difficulties with dualism are greatly enhanced by our modern evolutionary understanding of the origin of humanity from lower forms of animal life. If the mental is totally distinct from the physical, at what stage is its carrier, the soul, implanted in the physical body – if 'implanted' is the right word to use about an entity without extension? The issue had already arisen in connection with individuals, and the Fathers and the Scholastics had tended to assert that a foetus becomes 'ensouled' after some appropriate degree of development (of the order of forty days or more of embryological growth). We have to ask a similar question about the race itself. Did the Australopithecines participate in the mental or did that have to await the arrival of, say, *homo erectus*?

It is, of course, a logical possibility that God allowed the physical world to evolve until it contained subsystems of a certain degree of complexity and then conjoined to those subsystems – either once for all in a transmittable way, or individually as each developed – a mental component that we call the soul. If the mental and the physical are so separate, it is not clear why a high degree of complex physical structure is required before this happens. There seems to be an implied restraint on the nature of the physical before it can be a fitting receptacle for the mental, which hints at a connection between the two, unexplained by dualist theory.

The language has become explicitly theistic. There is a degree of inevitability about this appeal to deity as the Builder of the bridge between the worlds of mind and matter, for otherwise it is hard to see how they can be brought into any kind of realistic mutual relationship. In dualist thought

God becomes 'the God of the Gap' in a big way. Even Plato had to appeal to the demiurge to bring about the partial embodiment of the Forms in the brute matter (*hylē*) of this world.

One might think this appeal to theism would be congenial to a religious believer. I do not find it so, for it encourages a certain kind of supernaturalism which I do not take to be a fitting account of God's relationship with his creation. The main discussion of this point must await later chapters, but in essence I believe that the doctrine of creation and the demands of theodicy require a more genuine freedom for the world to be and to make itself than can be accommodated within a scheme that requires such radical and repeated intervention by the Creator in what is going on in his creation. In a dual world, the efficacy of the mental seems to require continual supernaturalistic support.

Gabriel Daly calls supernaturalism 'the mirror image of reductionism', and he says that it 'is not interested in, and may indeed be embarrassed by, the biological process by which our species came to be'.[41] Although dualism seems even-handed, in fact it almost inevitably enhances the mental at the expense of the material. Descartes believed that it was certain he could exist without his body, and on those sort of terms it is unclear what role of real value is fulfilled by the physical. In a Cartesian universe the animals are degraded to being mere machines, a picture at odds with the great appeals to the worth of the natural world expressed in biblical passages such as Job 38–41 and Psalm 104.

Among modern scientists there are few supporters of dualism. A notable exception is Sir John Eccles.[42] He even believes that he can use his great knowledge of neurophysiology to identify those sites in the 'liaison brain' where the interface between mind and matter takes place. The idea is to appeal to the physically indeterminate role of quantum events in certain delicate neuronal discharges, in order to provide room for mental manoeuvre. There are many disputable technical points about this proposal which are beyond my capacity to go into. It is interesting to note, however, that the theoretical physicist Roger Penrose has also made an appeal (in a very different way) to quantum theory as the basis for understanding 'the physics of the mind'.[43] His speculations depend upon an ingenious, but highly contentious, proposal that the solution of the measurement problem in quantum theory (what determines the outcome of a particular quantum event[44]) is related to the solution of the problem of quantum gravity, with

[41] Daly (1988), p. 132.
[42] Eccles (1984); (1989); Eccles and Robinson (1985).
[43] Penrose (1989), chs 9 and 10; see also Hodgson (1991); Lockwood (1989).
[44] Polkinghorne (1984), ch. 4; (1991), ch. 7.

the consequence that the behaviour of small systems retains what one might call a 'quantum plasticity' until their size reaches the Planck mass (10^{-5}gm, an enormously big value in atomic terms). I am not persuaded by these proposals. In an odd sort of way, the appeal to quantum theory – to the influence of mind in the outcome of quantum events – is a kind of microscopic occasionalism in which what happens in the world gets settled moment by moment.[45]

James Ross believes that 'Evolutionary hypotheses invite us away from the baleful ambiguities of dualism, from talk of our bodies and souls as if they were two things, with the soul the true self, the body the disposable shell.'[46] When one reads Eccles, one often gets the feeling that one of his motivations for believing in dualism is that he feels it is the only way in which to safeguard the reality of the mental and spiritual from the acid attack of a physicalist reductionism. Fortunately, I believe there is another approach which can achieve that desirable end. It is time to look at the second of the alternative strategies available to us.

One could summarize this as dual-aspect monism. There is only one stuff in the world (not two – the material and the mental), but it can occur in two contrasting states (material and mental phases, a physicist would say) which explain our perception of the difference between mind and matter. The attraction of the proposal is its seamless metaphysic, the continuity it maintains between the universe as quark soup and the universe as the home of self-conscious beings. The problem is to see what sense one could make of the proposal. The difficulties are considerable. Thomas Nagel says that 'talk about a dual aspect theory is largely hand waving. It is only to say roughly where the truth might be located, not what it is.'[47] Yet he is convinced that the quest for such an understanding should be undertaken. Under the rubric of the 'Possibility of Progress', he writes:

> The strange truth seems to be that certain complex, biologically generated physical systems, of which each of us is an example, have rich nonphysical properties. An integrated theory of reality must account for this, and I believe that if and when it arrives, probably not for centuries, it will alter our conception of the universe as radically as anything has to date.[48]

Let us consider what pre-Socratic flailings around have been taking place in this direction.

[45] The same would hold for Pollard's theory that God might act providentially in this way: Pollard (1958).
[46] In McMullin (1985), p. 223.
[47] Nagel (1986), p. 30.
[48] ibid., p. 51.

Twentieth-century philosophy has tended to eschew grand metaphysical syntheses in favour of a finely particular, not to say mousy, nibbling away at detail. Yet in A. N. Whitehead it has produced its one boldly creative thinker prepared to tackle great issues. His process philosophy[49] was framed as a proposal to describe the world in organismic rather than mechanical terms. Its basic category is that of event; what we consider as continuing entities are concatenations of events ('actual occasions'). Each event is bipolar. There is a prehensive phase, in which the individual event is linked to all past events, 'presented' with a portfolio of future possibilities, and influenced by a divine 'lure' to go in the best direction. However, initiative lies with the event itself; what actually happens depends upon the concrescent phase, in which the 'selected' option is realized. It is difficult to write about the notion without the extensive use of quotation marks, because of the deliberately volitional language which is used to express the flexibility built into this picture of process. Since an event could involve just a single elementary particle, this manner of speaking is problematic. It is not claimed, of course, that electrons make *conscious* choices, but in some way their behaviour is assimilated to ours. David Griffin, a notable contemporary exponent of process thought, says that 'the difference between the proton and the psyche is one of degree, not of kind (in an ontological sense). One who holds otherwise is a dualist, however odious such a description may be.'[50] Griffin does not accept the label panpsychic but prefers that Whitehead's metaphysic should be described as panexperiential. It might be said in reply that one who holds that the difference between the proton and the psyche is one of degree, not of kind, is a panpsychist, however odious such a description may be.

The way Whitehead seeks to avoid dualism is in fact by embedding it within the event itself, the quantum of process. Each of his events weds the mental (prehension) with the material (concrescence) in the marriage bed of occurrence. It is a kind of internalized dualism. But is it credible?

There are two major difficulties. One is its episodic character. David Pailin comments that 'The process of reality is thus treated as series of discrete instances. It jerks along rather than flows.'[51] It is sometimes thought that this is consonant with the fitful jerkiness of quantum theory, but that is not really so. Only certain special kinds of events (measurements) are discontinuous in conventional quantum mechanics. I do not believe that the insights of modern physics are particularly hospitable to Whitehead's views. It may be significant that it was in 1924 that he left the applied mathematics department at Imperial College, London, to take up a chair of

---

[49] Whitehead (1978).
[50] In Griffin (1986), p. 14.
[51] Pailin (1989), p. 53.

philosophy at Harvard. Thus the period in which he would have been in closest touch with the thinking of physicists came to an end just before the *anni mirabiles* of 1925–6 in which modern quantum theory came to birth.

The second difficulty is the panpsychic property attributed to matter. There are, of course, gradations – a stone is conceived as a mere aggregation of low-grade events, while a living cell is an integrated event of a higher category – but the language used shows that human experience is being employed as the clue to all experience, to a degree which fails to convince. Pailin, who is sympathetic to many aspects of process thought, concludes that 'on analysis the psychical description of reality as a whole seems either to be meaningless, misleading or deeply obscure'.[52] Ian Barbour has also expressed his reservations, for he feels that the Whiteheadian account is inadequate to express the *continuing* identity of the human self, and that it pays too little attention to emergence of new properties with increasing levels of complexity.[53]

There are aspects of Whitehead's thought from which one can benefit without accepting it in its entirety. Despite the fragmentary character resulting from its emphasis on discrete events, the prehensive poles of these events present a picture of a degree of interrelatedness which points one in a holistic direction. (This is in contrast with Leibniz's windowless monads, with which otherwise Whitehead's scheme has much in common.) There is a corresponding emphasis upon relationships as partly constituting entities, so that 'The atoms in a cell behave differently from the atoms in a stone. The cells in a brain behave differently from the cells in a plant.'[54] This provides an openness to the possibility of 'downward causation'; a whole, by its provision of context, can influence the behaviour of its parts. Later I will explore another way in which one might think of such downward causality.

Someone who is fully acquainted with modern physics, and who has made his own metaphysical proposals, is David Bohm.[55] Naturally, he is influenced by his own highly original interpretation of quantum theory, though it is not one which has commended itself to many other physicists. Bohm divorces wave and particle, which conventional quantum theory decrees to be complementary aspects of a single entity. The particles are what we see, and they are as unproblematically objective as any Newtonian might desire. The wave is unseen, but it guides the motion of the particles and encodes information about the whole environment in which they move. These notions have been expanded by Bohm into two generalized

---

[52] ibid., p. 148.
[53] Barbour (1990), p. 227.
[54] ibid., p. 228.
[55] Bohm (1980), especially ch. 7.

metaphysical concepts: the explicate order (particulate, manifest, the concern of a methodologically reductionist science) and the implicate order (holistic, veiled, the antidote to a merely reductionist view of reality). The explicate is enfolded in the implicate, which forms its ground. Consciousness is an aspect of the implicate order but it also has its explicate manifestation in the instant perceptions of the ego and in memory, just as the particles of matter arise from the same fundamental source. There is, therefore, a veiled interaction between mind and matter at the implicate level which enables causality to act in both directions. It is an entrancing picture, and Bohm has become something of a metaphysical guru with a quite extensive following, particularly in the United States.[56]

My instinct as a bottom-up thinker is to be wary of such grandiosities of philosophical fancy. Instead, I would want to follow the flight of such straws in a metaphysical wind as our understanding of the physical world provides. My own tentative ideas have been woven round two concepts: complementarity and openness. I have sought to explore their potentialities in other writings,[57] so I shall be almost as summary with myself as I have been with my distinguished predecessors.

When one thinks about the mental and the material in human experience one is struck both by their inextricable interrelation (the action of the will and the action of drugs) and by their qualitative distance from each other (the immense gap between neural activity and perceiving a patch of pink, let alone any higher form of mental activity). Such intimacy-in-opposition is strongly reminiscent of the quantum phenomenon of complementarity – for example, the wave/particle duality enjoyed by all quantum entities. One of the attractions of complementarity, when legitimately invoked, is that it succeeds in linking together concepts which otherwise would seem to be categorically disjoint (such as spread-out waves and point-like particles). Analysis shows that the way in which light can square the classical circle and possess both particle-like and wave-like behaviours lies in the presence of a quantum fuzziness within its structures. A wave-state contains an *indefinite* number of photons – an inconceivable possibility in the clear hard world of classical physics, where you have to have exactly a specific number of particles, no more nor less, but a possibility readily conceivable in quantum theory, whose superposition principle permits us to mix together states (like different numbers of photons) in a way rigidly forbidden in a common-sense Newtonian world.[58] A dual-aspect monism will be some sort of complementary mind/matter metaphysic (there is only one stuff, just as there is only

[56] See Weber (1980).
[57] Polkinghorne (1988), ch. 5; (1991), ch. 3.
[58] See Polkinghorne (1991), p. 86.

one light); and it seems possible, by analogy, that the clue to its consistency will lie in some radical indefiniteness present in the structure of the basic stuff.

It is possible that the direction in which to look for this indefiniteness is also hinted at by modern physics, for the twentieth century has seen the demise of a merely mechanical view of the universe.[59] To some extent this is due to quantum theory's cloudy fitfulness at the subatomic roots of the world. Yet I am hesitant to place too great a reliance on this factor. Partly my reluctance is due to the unsolved problems in the interpretation of quantum theory.[60] Different resolutions of these perplexities would yield greatly differing metaphysical consequences. Partly it is due to the fact that often microscopic uncertainties accumulate in ways that cancel each other out to produce macroscopic certainty. This is not always so, and there can be large-scale quantum consequences (superconducting effects are a case in point), but my instinct is to doubt whether the role of quantum theory is more than a small part of the story.

Much more significant, it seems to me, are the results from the study of exquisitely sensitive dynamical systems, which are gathered together under the (rather inapt) title of the theory of chaos.[61] Even in a Newtonian world, it seems, there would be many more clouds than clocks. Tame systems – in which small errors or ignorances about their present state produce only small consequent deviations in their expected future behaviour – are decidedly exceptional. Most of what we have to deal with in macroscopic physics is intrinsically unpredictable. Yet the future options are contained within certain limits (strange attractors). The theory of chaos gives a picture of behaviour which has the character of a kind of structured randomness, an oxymoronic ordered-disorder.

Unpredictability is an epistemological statement about what we can know. I do not think anyone would quarrel with it as a necessary ignorance extensively present in our attempted knowledge of the physical world. It is perfectly possible – and I have argued that it is natural and attractive – to go on to use it as the basis for an ontological conjecture. This asserts that actual physical reality is subtle and supple in its character; that physical process is open to the future. We live in a world of true becoming. Read from the bottom-upwards, physics provides us with no more than an envelope of possibility, within which future development is constrained to lie. Within

[59] I do not think that an important part was played in this by the development of classical field theory, since fields are essentially mechanical; see ibid., p. 93.
[60] ibid., pp. 89–92.
[61] See Gleick (1988); Stewart (1989); also my article 'The Laws of Physics and the Laws of Nature', in Russell et al. (1993).

that envelope, the path actually taken depends upon the realization of a specific set of options selected from among proliferating possibilities. These different possibilities are not discriminated from each other by energetic considerations (for otherwise the more energetically demanding trajectories would be excluded), but by something much more like an information-input (this path rather than that one). One sees the opportunity for using this information-input, necessary to resolve what actually occurs, as the vehicle for a downward operating causality, a role for the 'mental' (information) in the determination of the material. An epistemological defect has turned into a metaphysical opportunity. *Something* brings about the future, and if the reductionist account of physics does not suffice, room is thereby provided for the operation of further causal principles. The apparently deterministic equations, from which a bottom-up account of the classical theory of chaos begins, are then deemed to be *emergent-downward* approximations to the true, supple, physical reality. The approximation involved is probably that of treating constituents as isolable, for exquisite sensitivity implies that the smallest trigger from the environment can have large effects, so that there is an essential holism built into the nature of chaotic dynamics.[62] The possible significance of this for arguments about reductionism is discussed in a note at the end of the chapter.

The ontological picture proposed is one of increasing complexity generating increasing openness within which there is increasing scope for the use of explanatory causative concepts of a holistic and increasingly mental-looking kind. The open future of a world of becoming signifies that there are opportunities for the action of causal principles, other than the merely mechanical interaction of parts, in bringing that future about. Of course, immense difficulties remain unresolved. I present this point of view simply as indicating a hopeful direction in which to look. There is a considerable ontological distance between clouds and cells, and again between cells and people. Information and its processing are not the same as thought. I am persuaded by Roger Penrose's arguments of the fallacy of equating the two.[63] He makes the distinction between thinking (what people do) and executing algorithms (what computers do) by reference to the experience of doing mathematics. The proof of Gödel's theorem proceeds by constructing a statement which one can 'see' to be true, though it is also logically undemonstrable within the formalized system of argument under discussion. Polanyi put it with admirable succinctness many years ago:

---

[62] When considering small triggers, one should remember that there are some subtle and unresolved problems about the interrelation of chaos theory and quantum theory; see Davies (1989), pp. 365–70.

[63] Penrose (1989); see also Hodgson (1991).

'We know more than we can tell.'[64] It seems unlikely to me that our escape from the Gödelian trap is due, as Arbib and Hesse have suggested, to our being algorithmic machines but of a 'mistake-making' variety because we have to deliver an answer, willy-nilly, in a finite time.[65] Surely intellectual shots in the dark are no more the basis of fruitful thought than mere randomness would be the basis of freedom.

Penrose is a Platonist about the world of mathematical truth which we explore (as I have already confessed myself to be), but he recognizes also the specificities of our embodiment. Hence his concern with the 'physics of the mind' and his rejection of a strong AI (artificial intelligence) programme's claim that thought is software which could be run on any hardware. There may well be some things indispensably necessary in the neurophysiological structure of the brain that are not capable of being equivalently modelled in some vast array of electronic gadgetry. We are not just 'computers made of meat', and perhaps the meat plays an essential role in ensuring that this is so.

All this leaves the fundamental problem of consciousness still deeply mysterious. If we are looking in the right direction, it must lie at the far complementary pole, away out of sight from our bottom-up starting-point in chaotic dynamics (though always accessible to our top-down experience). It may well be that Nagel is right and it will be centuries before we can peer over that intellectual horizon. In the meantime we have a rudimentary picture of a physical world evolving entities like ourselves whose open flexibility enables them to participate in an everlasting noetic world of thought. (I am using mental in a wide sense that would comprehend the spiritual.) Out of the primeval quark soup have emerged saints and mathematicians.

I am not unduly perturbed by the modest results of our inquiry. The first lesson that a scientist learns is to take critically-assessed experience with the utmost seriousness, even when we are baffled by how to make integrated sense of it. Kamerlingh Onnes discovered the totally unsuspected property of superconductivity in 1911. More than fifty years elapsed before it was explained. It could not have been understood in 1911, since it is an intrinsically quantum mechanical phenomenon and modern quantum theory was then unknown. It would have been foolish to have taken its mysterious character as a reason for denying its existence.

It was always foolish to deny our basic experiences of consciousness and free agency on the grounds that they had no obvious place in a mechanical universe. We should be glad if physics begins to look a bit more hospitable in these matters, but we do not need the theory of chaotic dynamics to assure

[64] Polanyi (1958), *passim*.
[65] Arbib and Hesse (1986).

27

us that we face an open future. Holmes Rolston rightly reminds us that when someone looks through a telescope 'the most significant thing in the known universe is still immediately behind the eyes of the astronomer'.[66]

A NOTE ON REDUCTIONISM

Our investigations of physical reality can largely be ordered into a hierarchy of sciences whose objects of inquiry manifest an increasing degree of complexity: physics, chemistry, biology, anthropology. Those of a reductionist frame of mind regard the 'higher' sciences as no more than elaborations on the fundamental themes of the 'lower'. In the end, all is physics. Among contemporary scientists, the biologists, flushed with the undoubted successes of molecular biology, are most prone to make such assertions. Francis Crick proclaims that 'The ultimate aim of the modern movement in biology is in fact to explain *all* biology in terms of physics and chemistry.'[67] Such a thorough-going reductionism is a kind of neo-mechanical view of reality, and since it is the 'mechanical' problems which get solved first in the development of a science (clocks are easier to understand than clouds), it is scarcely surprising the biologists are tempted, in the first generation of their quantitative success, to espouse such opinions. Physicists did the same in the eighteenth century, but they have accumulated more experience since then (for physics is an easier subject than biology) and its practitioners, in consequence, tend to be much more wary about making claims that all is within their narrow grasp.

Those who write on science and religion generally resist reductionism.[68] There are a variety of ways in which to do so. An important distinction is between the claim that higher-level *theories* are autonomous because they involve irreducible concepts, and the stronger claim that higher-level *processes* have their own autonomy. We can characterize these attitudes as weak and strong anti-reductionism respectively. To make the distinction clear, the weak anti-reductionist recognizes that 'wetness' is not a concept applicable to individual H_2O molecules, but supposes the property to arise in aggregations of trillions of such molecules by the operation of *precisely* the same forces and effects as operate in and between small numbers of these molecules. On wetness the weak reductionist may well be right, but the difficulty about this position in general is that its new concepts seem to arise simply by abstracting from or averaging over greater complexity (as when

---

[66] Rolston (1987), p. 66.
[67] Crick (1966), p. 10.
[68] Barbour (1990), ch. 6; Peacocke (1979), ch. 4; (1986), chs 1 and 2; Polkinghorne (1986), ch. 6.

'temperature' signifies the average molecular kinetic energy in a gas), and it is difficult to see that this is likely to be adequate for understanding the truly puzzling and interesting emergences (such as life or consciousness) which appear to exhibit a qualitative novelty in their appearance.[69]

On the other hand, the invocation of strong anti-reductionism to explain these latter emergences is usually condemned as 'vitalism'. If the latter is held to involve the external addition of a magic extra ingredient, one can understand fully why it is rejected as improbable. The metaphysical egalitarianism implied by the suggestion in this chapter of both downward and upward (ontological) emergence appears to offer the prospect of a way out of this dilemma. Without extrinsic breach of continuity, it assigns a genuine novelty to complex systems, *in whose context the behaviour of the parts is not just the sum of their isolated behaviours*. There is scope for holistic laws of nature to operate in a mode of downward causality, and the constituent laws of physics are recognized as asymptotic approximations to a more subtle and supple physical reality. This point of view could fittingly be called *contextualism*.

If our conjecture is correct that it is in exquisitely sensitive systems that these properties are manifested, then those systems which are complicated without being unpredictable will retain their quasi-mechanical behaviour and will prove explicable in terms of the sum of their parts. In this way, the successes of molecular biology in exhibiting the physical basis of genetics can be accepted, without believing the whole of biology to be accessible through a similar treatment. DNA is quasi-mechanical; living organisms need not be. After all, as we noted, we have been here before. To the eighteenth-century successors of Newton, it appeared that physics was mechanical, but the twentieth-century insights of the dynamical theory of chaos show that this is only very partially the case.

[69] See note 40.

2

~~~~~~

Knowledge

'We *believe* . . .'

Our concern is with the search for truth. A religious belief can do all sorts of things for us – it can sustain us in life and in the approach of death; it can provide a thread of meaning in what would otherwise be a labyrinth of inanity – but it cannot do these things with integrity unless it is founded on the truth. I have great sympathy with David Pailin when he says that 'Attempts to defend theism by ignoring the question of truth . . . are fundamentally atheistic. They worship human wishes rather than ultimate reality.'[1] Richard Rorty is kind enough to say that 'Even today, more honest, reliable, fair minded people get elected to the Royal Society than to, for example, the House of Commons.'[2] The religious believer wishes to be found in the company of honest inquirers and not of polemicists for a cause.

How we arrive at a true belief is a question which has exercised many minds ever since philosophical reflection first began. The Enlightenment is a watershed in intellectual history because it represents the decisive repudiation of mere appeal to authority as the ground for belief. Immanuel Kant defined enlightenment as man's release from his 'inability to make use of his understanding without direction from another'.[3] It was a kind of intellectual coming-of-age, and its continuing spirit is clearly to be discerned in the terms of Lord Gifford's will. This release from an entail to external authority had been enabled by the success of seventeenth-century science in constructing an account of the physical world which seemed to owe little to Aristotle or Ptolemy (though it might have found an antique patron in Archimedes). What was repudiated was not a debt to the past, but enthralment to it. No one – least of all a scientist – can start intellectual history from scratch. Even Sir Isaac Newton had to say that if he had seen further it was because he had stood on the shoulders of giants. Whatever our

[1] Pailin (1989), p. 7.
[2] In McMullin (1988), p. 70.
[3] Quoted in ibid., p. 154.

intellectual discipline may be, we are heirs to its tradition, and though our generation may transform the understanding it inherits, it will do so on the basis of correcting the past rather than denying it. We do well to feel a profound suspicion of any claim that until now all has been darkness and only in our day has light begun to dawn.[4] Particularly is such caution called for in those subjects of inquiry where, unlike in science, we do not achieve a steady accumulation of agreed results. (I have suggested that the largely non-cumulative character of theology is a reflection of the fact that its Subject transcends us, while in the scientific exploration of the physical world we transcend the object of our investigation.[5]) The English mystics of the fourteenth century, and the Spanish mystics of the sixteenth century, may well have known things which we shall only learn by apprenticing ourselves to their insights.[6]

While the search for truth requires a critical evaluation of the past (and present), it is not likely to be assisted by a negative scepticism. The risk of initial commitment to what appears to be the case is a necessary part of finding out what is actually the case. Richard Swinburne has encapsulated this stance in the principles of credulity (other things being equal, probably things are as they seem) and testimony (other things being equal, probably things are as they are reported). 'The rational man is the credulous man – who trusts experience until it is found to mislead him – rather than the sceptic, who refuses to trust experience until it is found not to mislead him.'[7] Of course, the critical question is how to interpret the *ceteris paribus* clauses. There are many examples in the history of science of people refusing to recognize well-reported phenomena which did not fit in with their a priori point of view. Laplace (the greatest of Newton's successors in working out celestial mechanics) dismissed tales of meteors with the words, 'We've had enough of such myths.'[8] Yet I would not be disposed to take at face value a neighbour's description of how a flying saucer hovered over his garden the other evening. Credibility requires a continual interplay between the circumstantiality of a particular account and its relation to a general understanding of what is conceivable. (We shall face the problem in an acute form in a later chapter when I consider the resurrection of Christ.) I do not think there are formalizable rules to help us in our judgement. We simply have to be aware of the dangers of credulity and the dangers of scepticism, and pilot a path between them.

[4] For a history of scientific ideas sensitively appreciative of the insights of the past, see Park (1988).
[5] Polkinghorne (1991), p. 8.
[6] For example, see Jantzen (1987); Burrows (1987); Williams (1991).
[7] Swinburne (1979), p. 13.
[8] Quoted in Jaki (1989a), p. 97.

The search for truth is an intellectual adventure rather than the execution of a programmed procedure. I have argued elsewhere that the success of science should encourage us to be optimistic in general about the quest for knowledge, to bet upon our power to make verisimilitudinous sense of experience in many realms (cf. the note at the end of this chapter). Nor should we be dismayed if we detect that a certain circularity is involved in our thinking.[9] We have too long been bewitched by Euclid. A linear view of knowledge, as if it arose from building upon an unchallengeable foundation, does not work even in mathematics, as the nineteenth-century discovery of alternative geometries, and the twentieth-century recognition of the Gödelian incompleteness of axiomatized systems, make only too clear.[10] Instead, we are led to the circular view of knowledge which I have called an 'intellectual bootstrap', the self-sustaining character of a world-view, delivered from merely arbitrary concoction by being required to submit to the rational criteria of comprehensiveness and economy.

At least two circularities are involved in the search for knowledge. One is the hermeneutic circle: we have to believe in order to understand and we have to understand in order to believe. That is why scepticism is so unfruitful a strategy. Why do I believe in quarks when no fractionally charged particle has ever unequivocally been observed in an experiment? Set your doubts aside for a while and see how belief in confined quarks enables us to understand a variety of phenomena (the hadronic spectrum of octets and decuplets; deep inelastic scatterings[11]) which otherwise would have no underlying intelligibility. Yet I cannot go on forever invoking the unexplained hypothesis of quark confinement, and I must hope that one day we shall understand the theory of quark behaviour (quantum chromodynamics) to the extent that we shall see that confinement is entailed by it.

The second circle is the epistemic circle: how we know is controlled by the nature of the object and the nature of the object is revealed through our knowledge of it. There must be a mutual conformity between the act of knowing and the known, and our concepts must arise in a way that respects that conformity. Our access to the quantum world is restricted by the limits of Heisenberg's uncertainty principle. If we demand the localized clarities of a Newtonian epistemology, we shall deny ourselves all knowledge of quantum entities. If we accept the veiling of that world as part of its nature, then we shall encounter its idiosyncratic reality, even if it will occasion us a good deal of ontological perplexity.[12] It can only be met on its own terms.

[9] Polkinghorne (1991), ch. 1.
[10] cf. Puddefoot (1987).
[11] See Polkinghorne (1989b).
[12] d'Espagnat (1989); Polkinghorne (1991), ch. 7.

We cannot impose a rigid grid of expectation on the physical world, but we must allow its facticity to mould our thought and ways of knowing it. Thomas Torrance has been particularly emphatic in stressing the necessarily open dependence of knowledge on the nature of that which is known. He calls for 'the cultivation of what Clerk Maxwell called a *new mathesis*, through which we allow fresh revelations of nature to call forth from our minds new modes of thought by which the process of our minds can be brought into the most complete harmony with the process of nature'.[13] Equally, theologians must operate 'in accordance with the fluid axiomatic modes of thought that have served so well to reveal the mysteries of nature'.[14] 'Theological science must therefore generate concepts of God that are *worthy* of him, as the great Origen used to insist . . . concepts that are forced on our minds by the sheer nature of the divine Majesty.'[15]

> We cannot begin by forming independently a theory of how God is knowable and then seek to test it out or indeed to actualize it and fill it with material content. How God can be known must be determined from first to last by the way in which He actually is known.[16]

For a scientist with a serious interest in theology, this emphasis on the epistemic primacy of the object of knowledge is one of the most valuable and congenial aspects of Torrance's thought. One is only too aware of how inadequate our powers of rational prevision are to anticipate the surprising way the world is. I have already referred to quantum theory as an illustration of the need to allow our thinking to be shaped by the nature of the reality encountered.

God is known because he has chosen to make himself known, through gracious disclosure. This revelatory action does not take the form of a mysterious conveyance of incontestable propositional knowledge; rather, it is mediated through events and people which have the character of a particular transparency to the divine presence and to intimations of a lasting hope.[17] Ronald Thiemann has suggested that the doctrine of revelation 'can be given self-consistent formulation if it is organized around the notion of "narrated promise" '.[18] I would wish to emphasize historical anchorage by replacing that phrase with '*enacted* promise'. A bottom-up thinker does not wish simply to tell a story, but to point to occurrence. Thiemann says that

---

[13] Torrance (1985), p. 79.
[14] ibid., p. 81.
[15] ibid., p. 83.
[16] Torrance (1969), p. 9.
[17] Polkinghorne (1991), ch. 4. I regard worship and hope as fundamental aspects of religious experience.
[18] Thiemann (1985), p. 96.

'To acknowledge the biblical narrative as God's promise is to believe that the crucified Jesus lives,' but he adds, 'Theology can explain neither why nor how persons come to believe such a paradoxical claim.'[19] I do not pretend that belief in the resurrection is demonstrable beyond a peradventure, but in chapter 6 I shall seek to show that it is rationally motivated.

How does it all work out in practice? Thomas Huxley counselled us to 'sit down before fact as a little child, prepared to give up every preconceived notion, follow humbly wherever and to whatever abysses nature leads you, or you shall learn nothing'.[20] That advice would not surprise the writer of the First Epistle of John, who expresses his theological epistemology in the words:

> That which was from the beginning, which we have heard, which we have seen with our eyes, which we have looked upon and touched with our hands, concerning the word of life – the life was made manifest, and we saw it, and testify to it, and proclaim to you the eternal life which was with the Father and was made manifest to us – that which we have seen and heard we proclaim also to you . . . (1 John 1.1–3)

But, of course, we have grown more sophisticated and we know that the concept of 'fact' is far from being unproblematic. Scientific facts are not uncontroversial matters, like electronic counter readings or marks on photographic plates, but they are the interpretations of those raw registrations, interpretations which are themselves embedded deep in current theoretical understanding, so that those readings become the signal of a Z° and those marks the signs of the decay of a triply strange Ω^-. There is a symbiosis between theory and experiment; we cannot survey the world without donning 'spectacles behind the eyes'.[21]

The more deeply personal the encounter with reality, the more profoundly will its significance depend upon the interpretation attributed to it by its participants. Or should one say, the more vulnerable it will become to the cultural shaping of those mental spectacles? 'There was again a division among the Jews because of these words. Many of them said, "He has a demon, and he is mad; why listen to him?" Others said, "These are not the sayings of one who has a demon. Can a demon open the eyes of the blind?" ' (John 10.19–21).

The diversity of religious claims can be seen as a cancelling cacophony. Or it can be seen as the inevitable consequence of the search for One whose glory must be veiled, whose infinitude can never be caught in our finite

[19] ibid., p. 147.
[20] Quoted in Jaki (1989a), p. 105.
[21] Hanson (1969), ch. 9.

nets, whose light is refracted by the cultural prisms of humanity. The matter cannot be settled by a priori argument. While there is a theological tradition which seeks, in the words of Vincent of Lérins, to appeal to what has been believed 'everywhere and at all times by all people', there is also an understanding which recognizes that controversy is endemic in the life of the Church, so that 'any realistic account of the Christian phenomenon strongly suggests the inconceivability of there ever being complete agreement about the identity of Christianity. That is not to say that Christians may not be able to contain disagreement within reasonable boundaries. But contained diversity, is, in fact, what unity amounts to.'[22] There is not a standard prescription for the Christian spectacles behind the eyes. Nancey Murphy takes a cheerful view of such variety, speaking of the different denominations and sects as representing possibilities for 'laboratories for theological experimentation'.[23] One could regard their differing perspectives as contributing to a stereoscopic whole, so that one need not strive for ecclesiastical uniformity, but rather a diversity of insights. However, our perplexity is bound to increase when we go beyond the Christian fold to consider the bewildering varieties of world-wide religious traditions. (That is a subject for a later chapter.) I am certain, however, that our search for knowledge of God will have to seek an anchorage in experience; that theology stands in need of data which in George Tyrrell's words are 'not tacked down to the table by religious authority',[24] but are open to assessment. It is because I regard the Nicene Creed as being a condensed 'data table' of that kind that I shall be exploring its statements in the chapters which follow.

The data for Christian theology are to be found in scripture and the tradition of the Church (including, of course, the contribution of our own experience), and in such general insights about order and purpose that may be brought to light by the play of reason on the process of the world. Together, these constitute what a physicist would characterize as a 'phenomenological' account – that is to say, an array of data arranged in suggestive patterns. In physics one would then want to go on to try to form a theory – that is to say, an economical interpretation of wide empirical adequacy which would be a candidate for the verisimilitudinous description of what was actually happening. In science, the transition from phenomenology to theory would be exemplified by the move from just talking of particles as made up of quarks to talk of a basic field theory for the understanding of quark behaviour (quantum chromodynamics). In theology,

[22] Sykes (1984), p. 11.
[23] Murphy (1990), p. 166.
[24] Quoted in ibid., p. 89.

the transition from phenomenology to theory would be something like moving from the basic Christian assertion that 'Jesus is Lord' to a doctrine of the incarnation. I have elsewhere given reasons for believing that the task of theology is so difficult that it is unreasonable to expect the fully successful accomplishment of theory construction.[25] Instead, one may have to settle for a portfolio of different models – none claiming an exhaustive correspondence with the way things are, but each usable with discretion – in order to cast light upon an appropriate range of phenomena.

I do not think we should be discouraged by the realistic modesty of these remarks. Even in science, when we move away from the comparative simplicities of elementary particle physics into the complexities of condensed matter theory or the biology of organisms, we have to settle for whatever intellectual gains we can get. Quantum theory exhibits great explanatory power (we can use it to discuss the behaviour of quarks inside protons or the nuclear processes inside stars), but our unresolved perplexities about the measurement problem[26] show that we still do not *understand* it after more than sixty years of exploration. By the same token, perhaps it is not surprising that despite many attempts to explain the problem of evil,[27] its intransigent mystery defies our theological understanding.

The distinction between explanation and understanding is very important for theology. Understanding in science is a deep experience going beyond mere predictive power or the currently fashionable notion of algorithmic compressibility.[28] The latter operates when a simple rule can be used to generate complicated consequences. To see that neither is equivalent to 'understanding', consider what would happen if the weather were found to be governed by the law that it was always the same on the same day of the same month, whatever the calendar year. Such a rule would be powerfully predictive and highly compact. But its discovery would not have quenched our thirst for understanding. *Why* should the weather exhibit this remarkable cyclicity? Only when a satisfying answer had been given to that question would we have felt that we understood the matter.

To understand something is to feel an intellectual contentment with the picture being entertained. Sometimes that will come about by the patient sifting and consolidation of proffered explanation. The so-called Standard Model for the structure of matter[29] arose in this way, through thirty years

[25] Polkinghorne (1991), ch. 2.
[26] ibid., ch. 7.
[27] Hick (1966b); see also Polkinghorne (1989a), ch. 5.
[28] See Barrow (1992); Davies (1992), ch. 5.
[29] Polkinghorne (1989b).

of endeavour by elementary particle physicists. The work of the early ecumenical councils from Nicaea to Chalcedon offers some analogy in Christian theology. Sometimes understanding will come about by the convergence of many lines of argument upon a common conclusion – the process which William Whewell called 'consilience'. Historico-scientific theories (such as big-bang cosmology or evolutionary biology) usually establish themselves in this way. The cumulative case for belief in God[30] would be a theological instance. Sometimes understanding is gained through the adoption of a new perspective from whose aspect previously disorderly facts are seen to fall into a novel and satisfying pattern. The recognition of the helical structure of DNA transformed attempts to understand the molecular basis of genetics.[31] The Christian concept of a suffering Messiah made sense of Jesus' relation to Israel, which could not be achieved on the basis of the expectation of a military deliverer.

It is possible for understanding to be attained without the possession of a detailed explanation. Darwin felt he understood the evolving complexity of terrestrial life, though his ignorance of Mendel's genetic discoveries meant that he could not explain the source of the variations on which natural selection had to work. In Christian theology, the saving power of the death of Christ had been understood from the first centuries of the Church's existence, though theories of the atonement arrived late in its thinking and they have proved only partially explanatory.

The ability of understanding to outrun explanation is intimately connected with the religious concept of faith. This is not a polite expression for unsubstantiated assertion, but it points to an ability to grasp things in totality, the occurrence of an insight which is satisfying to the point of being self-authenticating, without dependence on detailed analysis. Involved is a leap of the mind – not into the dark, but into the light. The attainment of understanding in this way does not remove the obligation to seek subsequent explanation, to the degree that it is available, but the insight brings with it a tacit assurance that such explanation should be there for the eventual finding. Such experiences are quite common among scientists. Paul Dirac tells us how one of his foundational ideas about quantum theory came to him 'in a flash' when he was out for a Sunday walk. He was too cautious to be sure immediately that it was right, and the fact that the libraries were closed prevented his checking it right away. Nevertheless, 'confidence gradually grew in the course of the night', and Monday morning showed that his idea was indeed sound.[32] The mathematician Henri Poincaré was more certain

---

[30] Mitchell (1973); Swinburne (1979).
[31] Crick (1988).
[32] Kragh (1990), p. 17.

of his insight. An important idea came to him 'At the moment I put my foot on the step [of a bus] . . . I did not verify the idea . . . but I felt a perfect certainty.'[33]

Not all illuminations of faith come in this peremptory Damascus-road manner; many will involve a growing conviction whose coming to maturity may not be datable. Their common essence is the attainment of understanding by the power of a whole, by intuitive grasp rather than detailed argument. Torrance speaks of

> the distinction between the discursive mode of demonstration found in geometry and kindred sciences . . . and the ontological mode of demonstration that arises when something utterly new becomes disclosed and our minds cannot but yield conceptual assent to its self-evident reality. Such an act of assent was also spoken of as the response of *faith* made in recognition and acknowledgement of a truth that seizes the mind and will not let it go. Genuine faith in God, for example, was held to involve a conceptual assent of this kind, as the human mind is allowed to respond faithfully to God's self-evidencing revelation.[34]

Involved in this movement of faith is the exercise of those tacit skills which Michael Polanyi rightly diagnosed as indispensable to the scientific enterprise and which give it kinship with all other forms of human rational inquiry.[35] 'We know more than we can tell.' In this way it is possible to transcend the limitations of logical system-building. Austin Farrer concluded an ingenious but not wholly convincing discussion of the perplexities of theodicy by acknowledging

> that the value of speculative answers, however judicious, is limited. They clear the way for an apprehension of truth which speculation alone is powerless to reach. Peasants and housekeepers find what philosophers seek in vain; the substance of truth is grasped not by argument, but by faith.[36]

Recognition of the limitations of ratiocination is not indulgence in anti-intellectualism, but rather the avowal that knowledge has a broader base than that afforded by atomized argument alone.

Understanding is the crown of knowledge, and its attainment calls for an intuitive synthesis of meaning. Pannenberg says that 'Hegel described this synthesis as speculative intuition which supplemented and transcended

[33] Quoted in Penrose (1989), p. 419.
[34] Torrance (1989), p. 74.
[35] Polanyi (1958).
[36] Farrer (1962), p. 187.

38

reflection.'[37] Because of its tacit character, such knowing cannot be reduced to the observance of epistemological rules. We should neither abuse, nor be dismayed by, this absence of a formalized methodology. Scientific method has proved equally elusive to philosophical specification, and Ernan McMullin wisely comments about attempts to tackle that problem that

> the crucial question is: do they illuminate what the scientist is actually *doing*? And the danger is that to the extent that they do not, this will be taken by the philosophers to be a deficiency on the part of the scientist instead of an inadequacy in his or her formalism.[38]

Because we cannot tell, it does not mean that we do not know. Polanyi said of those who refused to allow an intuitive exercise of scientific induction that they 'would reject the acknowledgement of such powers of discovery as mystery mongering. Yet those powers are not more mysterious than our powers of perception but of course not less mysterious, either.'[39] Thomas Nagel said of our capacity to form useful concepts beyond the limit of our immediate experience that 'We are supported in this aim by a kind of intellectual optimism: the belief that we possess an open-ended capacity for understanding what we have not yet conceived.'[40]

Theology, of course, faces an additional problem in that its method is not only elusive but its infinite Subject is also necessarily beyond the total grasp of finite minds. A consequence is that it is easier to say negatively what is not the case than to describe positively what is the case. Much theological thought has been provoked as a response to what is found to be unsatisfactory. The Chalcedonian definition is concerned with fencing off the unacceptable from the acceptable; it is more successful in rejecting the Christologically heretical than in articulating the Christologically orthodox. Langdon Gilkey says that 'Usually it takes the heretic to create the theologian – a fact which professional theologians should remember with more gratitude than is their wont.' He goes on to say, 'To understand a doctrine, therefore, we must first of all understand what it denies, and then seek to understand the deep positive affirmation that it hopes to preserve.'[41] In the language of the mathematical physicists, a good deal of theology is concerned with 'no-go theorems'.

It would be no bad thing at this stage to recall the warnings of apophatic theology against rational over-confidence in our prattling talk about the

[37] Pannenberg (1976), p. 340.
[38] McMullin (1988), p. 24.
[39] Polanyi (1969), p. 167.
[40] Nagel (1986), p. 24.
[41] Gilkey (1959), pp. 47–8.

mystery of deity. We need intellectual daring but not foolhardiness. I should temper my optimistic quotation from Nagel by citing his subsequent sentence. 'But we must also admit that the world probably reaches beyond our capacity to understand it, no matter how far we travel.'[42] How much more so must that be true of God.

Stanley Jaki may say with some impatience, 'if pointers do not point unambiguously, that is, with certainty, what is the point of using them?',[43] but God is not to be read out of experience with quite that degree of clarity. I have more sympathy with the words of David Burrell, who speaks of the aim 'to secure the distinction of God from the world, and to do so in such a way as to display how such a One, who must be unknowable, may also be known'. He says that 'the question lies at the margins of human understanding and so at the intersection of reason with faith – a locus one may well be aware of without being able to inhabit'.[44]

Attempts to articulate the knowledge of God will require language to be stretched by appeal to analogy. In consequence theology must avail itself of the openness of reference provided by symbol.[45] Such a recognition is the essence of a rational theology, not its negation. Recourse to symbol is in no way incompatible with the assertion of truth; indeed, some deep truths can only be expressed symbolically. Even our encounter with the mystery of human personality makes similar demands – hence the importance of the poets and artists. Arbib and Hesse say, 'It is not so much that symbolism violates logic as that it constitutes a more general system than logic and that non-logicality is pervasive in language itself as well as in non-verbal symbolism.'[46] A capacious understanding of the nature of rationality is called for if we are to do justice to the rich variety of experience. Ward says:

> The highest use of philosophical reason lies in the conceiving and application of a new organizing idea, or a new interpretation of an existing idea, which enables one to build up a new, more comprehensive scheme for understanding the world. That is a function of imaginative and creative reason. It is certainly not deductive, for that only works out what is already there. And it is not inductive either. It is a presuppositional activity, which picks out and organizes the primary data in a particular imaginative way; it is like constructing a pattern for the world to fit into, from the creative extension of a number of clues.[47]

[42] Nagel (1986), p. 24.
[43] Jaki (1989b), p. 221.
[44] Burrell (1986), pp. 3–4.
[45] Polkinghorne (1991), ch. 2.
[46] Arbib and Hesse (1986), p. 207.
[47] Ward (1982a), p. 110.

In common with many others, I have wished to revalue the classical 'proofs' of God's existence as suggestive insights rather than logically coercive demonstrations.[48] They are part of those consilient 'converging lines of probable reasoning'[49] which constitute a case for theism. The logical inappropriateness of any other point of view is emphasized by Gilkey: 'a god who is proved is reached indirectly through implications from other, more ultimate principles or certainties'.[50] (A modern example of how those principles might be held to play that role is given by John Leslie's appeal to 'creative ethical requiredness' to do the work traditionally assigned to a personal God.[51] In my view, Leslie's discussion represents the danger of pushing too forcefully the axiological 'proof', valuable as I find the latter in the more modest role of an insight.)

Central to all talk of the knowledge of God is the recognition that it is not available to us on a merely speculative basis. I believe in quarks, but the acknowledgement of their existence does not touch or threaten me in my own being. It is very different with belief in God, which has consequences for all that I do and hope for. 'Whence then comes wisdom? And where is the place of understanding? . . . "Behold the fear of the Lord, that is wisdom; and to depart from evil is understanding" ' (Job 28.20, 28). Leibniz said that 'If geometry were as much opposed to our passions and present interests as is ethics, we would contest it and violate it but little less, notwithstanding the demonstrations of Euclid.'[52] We have to be scrupulous to see that our search for religious truth is not similarly clouded by resistance to submission to God's will or by desire to console ourselves with a heavenly comfort blanket. On the other hand, the recognition of an inescapable entailment for conduct in the embrace of a religious belief does not mean that the latter is reducible to a symbolic statement of how we intend to behave. Expressivist views of religion are very popular today,[53] but I cannot renege on the commitment to the cognitive quest with which I began this chapter, for the way things are is the only reliable basis for the way we should respond to them.

One reason for the attraction of an expressivist view is that it provides an obvious explanation of what many see as the unreasoning resistance of religion to falsification. On this view, we have simply nailed our moral colours to the mast, and an uncompromising persistence in conduct is

[48] Polkinghorne (1988), ch. 1; cf. Ward (1982b), ch. 2.
[49] J. H. Newman, quoted in Burrell (1986), p. 6.
[50] Gilkey (1959), p. 441.
[51] Leslie (1989), ch. 8.
[52] Quoted in Jaki (1989a), p. 45.
[53] Cupitt (1980); Phillips (1976).

admirable in a way that an uncompromising persistence in belief may not be. In the one case we exhibit resolution; in the other we exhibit pig-headedness. I have elsewhere given my reasons for rejecting a merely obscurantist, incorrigible picture of religious belief.[54] Theology shares with any other metaphysical world-view that generality of account which means that it is neither impervious to contrary evidence nor immediately falsifiable by it. Michael Banner has given further arguments for not capitulating to settling for a merely expressivist description. He lays emphasis on the phenomenon of a loss of faith: 'An obvious feature of the religious life is the desertion of it . . . How does it happen that people fall away from religious faith?'[55] This is readily understandable on a cognitive basis (it is simply that revision of belief which was claimed not to occur, but which certainly does, both in loss of faith and in conversion); it is a much more problematic phenomenon on an expressivist view, particularly since it is often unaccompanied by any drastic revision of moral priorities.

Furthermore, expressivist views take too existential, psyche-centred, a view of religious belief. Natural theology (understood as the search for God through reason and general experience) is an antidote, reminding us that there are reasons for such belief lying outside the behaviour patterns of men and women. I do not want to rehearse again the considerations of natural beauty and fruitful balance, found in the basic structure of the cosmos, which have led in our time to a revival of natural theology as a route to belief in God.[56] Essentially this revival stems from the recognition that the physical universe is not satisfactorily self-explanatory, that science by itself will not quench our intellectual thirst for understanding, so that we must explore the possibility that for the contingent world, as Torrance said, 'its deepest secret lies outside its own reality'.[57] The impetus for this movement has come largely from the physicists. Theologians have been more reserved. Those in a Barthian tradition can regard the whole project as misconceived. 'The logic of the matter demands that, even if we only lend our little finger to natural theology, there necessarily follows the denial of the revelation of God in Jesus Christ. A natural theology which does not strive to be the only master is not a natural theology.'[58] That quotation from Karl Barth is taken from his commentary on the Barmen declaration against the Nazis, and I believe it is unduly influenced by his belief that 'German theology', produced by that part of the Church which was subservient to Hitler, was

---

[54] Polkinghorne (1991), ch. 4.
[55] Banner (1990), p. 89.
[56] Davies (1983); (1992); Montefiore (1985); Polkinghorne (1986); (1988).
[57] Torrance (1981), p. 36.
[58] Green (1989), p. 174.

the kind of perversion to which natural theology inevitably tends. The answer is, surely, not to deny natural theology, but to insist on its integration with the rest of theological discipline in a single endeavour to learn of God. This has been a persistent theme in the writings of Thomas Torrance,[59] despite the influence of Barth upon his thinking generally. Nancey Murphy puts it well when she says, 'Thus it appears that the crucial step . . . must be an approach to theology that does not distinguish between natural theology and theology of revelation, but one that draws upon religious tradition to provide a measured concept of God.'[60] She is commending this as a strategy for an 'effective apologetic', but I think it is equally necessary for effective systematic theology.

Not all suspicion of natural theology is so ideologically grounded as the Barthians'. Much of it arises from a feeling that theologians' fingers were burned when the eighteenth- and nineteenth-century physico-theology (of Paley and the Bridgewater treatises) crumbled as Darwin struck at the foundations of its argument from design, and that in future it is best to fear the fire. Certainly there are lessons to be learned from that episode. The modern revival of natural theology does not look to occurrences within the process of the world (it does not assert that only God's direct action could cause life to arise from inanimate materials), but it looks to the basic laws which are the ground of that process (e.g., the Anthropic Principle[61]). Here is the answer to Hume's criticism that design arguments for a Creator were too anthropomorphic, being based on the analogy of human construction. The appeal of the modern insight is to the fabric of the world, whose delicate balance is necessary for the very possibility that the universe can make itself through fruitful evolution. We do seem here, in Hebrew terms, to be concerned with *bārā* (divine causing to be), not *'āsāh* (human making), with fundamental nature rather than contingent history.

Mistakes by natural theologians in the past do not preclude the possibility of success in the present. The science of 1750–1850 made plenty of mistakes too (phlogiston, caloric), and had ideas which were eventually fruitful but in a somewhat different form from that envisioned by their originators (Prout's hypothesis, that elements are composed of multiples of hydrogen, was a forerunner of modern ideas of nuclear structure). As a scientist I am often struck by theologians' persistent fears of getting it wrong. I don't quite go along with Popper's notion of 'bold conjectures'[62] as the basis of doing

---

[59] Torrance (1969); (1985).

[60] Murphy (1990), p. 18.

[61] Barrow and Tipler (1986); Leslie (1989); see Polkinghorne (1991), ch. 6, for a critical discussion.

[62] Popper (1959); (1963).

science (the method is subject to more rational constraint than that; one does not just draw a bow at a refutable venture), but a willingness to explore ideas which might prove mistaken, or in need of revision, is a necessary price of scientific progress. One would have thought that the intrinsic difficulty of doing theology would encourage a similar intrepidity. At times (the patristic period, the Reformation) that has been so, but not always. I am not, of course, denying the existence of many wild flights of contemporary theological fancy, but saying that within its more sober core I detect a degree of disinclination to take intellectual risk, particularly where it involves interaction with another discipline. Hence the widespread neglect of natural science by theologians.

If natural theology is truly to flourish again, it will require more input from the theological side. It would also, I believe, benefit greatly from a more serious involvement by scientists with a biological background. They might be reluctant because memories of the Darwinian débâcle are still kept alive in their community, and are even stirred up afresh by the farce of 'creation science'. They might fear that the fitting god of biological natural theology would prove to be Shiva, Lord of creation and destruction. The latter thought should no more deter them than fear that the fitting god of physical natural theology would prove to be the impersonal abstraction of the Great Mathematician should deter the physicists. Both forms of inquiry are limited in their characters and consequently yield only limited insights, in need of correction and revision from other sources of understanding about the divine nature. Michael Buckley has shown persuasively how the eighteenth-century roots of modern atheism lay in a loss of nerve on the part of the theologians who relied simply on a kind of philosophical natural theology and set aside the central appeal to Christian experience.

> Religion abandoned the justification intrinsic to its own nature and experience, and insisted that its vindication would be found in philo-sophy, become natural philosophy, become mechanics . . . if religion itself has no inherent grounds on which to base its assertion, it is only a question of time until its inner emptiness emerges as positive denial.[63]

Theology without natural theology would be in a ghetto, cut off from knowledge of the physical creation; natural theology by itself would be vulnerable, apt to seem little more than a competing possibility alongside a thorough-going naturalism. Once again one sees how essential it is that theological inquiry is conducted as a fully integrated discipline.

There is also a kind of natural theology which can draw material from the

---

[63] Buckley (1987), pp. 359–60.

insights of the humanities. Jeremy Begbie quotes Tillich's dictum that 'As religion is the substance of culture, so culture is the form of religion,' and he comments that 'Every cultural act is therefore implicitly religious, even if not by intention. It is necessarily rooted in the unconditioned meaning or ground of all meaning.'[64] Maritain said that 'the artist, whether he knows it or not, is consulting God when he looks at things'.[65] The mystery of art and literature – that sounds and paint and words on a page are vehicles by which we gain access to what we perforce acknowledge as profound truth – is a fact of great significance. I cannot dismiss such encounters as epiphenomenal – pleasant poetic ripples on the surface of prosaic reality. Rather they are revelations, unveilings of a deeper meaning.[66] That the world is the carrier of such meaning seems most naturally understood if it is a creation shot through with intimations of its Creator. George Steiner says, 'I can only put it this way (and every true poem, piece of music or painting says it better): there is aesthetic creation because there is *creation*.'[67] Earlier he had said, 'There is language, there is art, because there is the "the other".'[68] Once again we engage with signals of transcendence. Of particular significance to Steiner is music's character of 'time made free of temporality'.[69] He says that he believes 'the matter of music to be central to that of the meanings of man, of man's access to or abstention from metaphysical experience'.[70] That a temporal succession of vibrations in the air can speak to us of eternity is a fact that must be accommodated in any adequate account of reality. I agree that 'The questions: "What is poetry, music, art?", "How can they not be?", "How do they act upon us and how do we interpret their action?" are, ultimately, theological questions,'[71] for I think that our aesthetic experience is to be understood as a sharing in the joy of creation and also in the pain of creation's birth and redemption (since we must heed Barth's warning that '*God's beauty* embraces death as well as life, fear as well as joy, what we might call ugly as well as what we might call beautiful'[72]). I want to say to my friends who have difficulty with religious belief, 'Can you deny music? If not, then you acknowledge a dimension of reality transcending the material, which may prove for you the intimation of what Steiner calls a "real

[64] Begbie (1991), p. 7.
[65] Quoted in Pattison (1991). This reference gives an account of various theological approaches to the visual arts.
[66] See the defence of art as knowledge by J. Begbie in Montefiore (1992), pp. 58–83.
[67] Steiner (1989), p. 201.
[68] ibid., p. 137.
[69] ibid., p. 27.
[70] ibid., p. 6.
[71] ibid., p. 227.
[72] Quoted in Begbie (1991), p. 224.

presence".' The natural theology of the arts is a subject of the highest importance.

In its literary mode it will explore the inwardness of human life as a clue to our creaturely nature. Howard Root wrote:

> Where do we look for fruitful, stimulating, profound accounts of what it is to be alive in the twentieth century? . . . We look to the poet or novelist or dramatist or film producer. In creative works of art we see ourselves anew, come to understand ourselves better and come into touch with just those sources of the imagination which should nourish efforts in natural theology.[73]

He goes on to say, 'The starting-point for natural theology is not argument but sharpened awareness.'[74] In a word, as I have already insisted, the character of natural theology is insightful rather than demonstrative. Its disclosures will have many sources. I agree that 'All art is about revelation,' but I believe that Timothy Gorringe displays his undue suspicion of cool rational inquiry rooted in human understanding when he goes on to say, 'art, in fact, is the only valid form of natural theology'.[75]

This chapter has been a variation on a familiar theme. Its discussion of theological knowledge has tried to do justice to the idiosyncrasy of the discipline, while at the same time assimilating it to many other forms of human rational inquiry, including science. This is a strategy popular among scientists with serious interests in theology,[76] and it is accepted by some theologians also.[77] Pannenberg speaks of theology as the 'Science of God', saying that this means 'the study of the totality of the real from the point of view of the reality which ultimately determines it both as a whole and in its parts'.[78] I see theology as having a dual role: firstly as the specialist investigation of particular types of experience and insight which we label religious (systematic theology); and secondly as the great integrating discipline which incorporates the results of all specialist investigations (including science's and its own) in a profoundly comprehensive matrix of understanding (philosophical theology). In its first role theology can be said to be one of the sciences; in its second role it is *the* metascience. Because I regard the existence of God as the ground for the possibility of this latter

[73] In Vidler (1962), p. 18.
[74] ibid., p. 19.
[75] Gorringe (1990), p. 121.
[76] Barbour (1974); (1990), chs 2 and 3; Carnes (1982); Peacocke (1984); Polkinghorne (1991), chs 1 and 2.
[77] Lonergan (1972); Pannenberg (1976); Torrance (1969); (1985).
[78] Pannenberg (1976), p. 303.

integrating discipline, I prefer to call it philosophical theology rather than metaphysics. The unity of knowledge and the unity of God are closely connected. I believe in both.

Belief that science and theology are intellectual cousins under the skin encourages theological interest in the philosophy of science. It will be apparent that I am considerably influenced by the writings of Michael Polanyi.[79] I find his account of what scientists are doing to be one which is actually recognizable by a practising scientist.[80] The emphasis on the exercise of tacit skills of judgement, within a convivial community but employed with universal intent, is both consonant with what goes on in science (which, of course, Polanyi, unlike most philosophers of science, actually knew from the inside) and it has obvious affinities with the procedures of other disciplines – like theology, where it is even harder to tell what we know and why we know it. Polanyi said that 'Comprehension is neither an arbitrary act nor a passive experience, but a responsible act claiming universal validity. Such knowing is indeed *objective* in the sense of establishing contact with a hidden reality.'[81]

Recently, another philosopher of science has been attracting a good deal of attention among theologians. He is Imre Lakatos.[82] His description of scientific method was developed in the attempt to give a more satisfactory account of progress in science than had been afforded by the views of Karl Popper.[83] The latter's well-known emphasis on the role of falsifiability had clearly placed a finger on a significant aspect of scientific procedure, but by itself it did not account well for the actual history of science. Low-level hypotheses are no doubt vulnerable to ready falsification (I predict that the red liquid will turn green when I add factor X – if it doesn't, too bad for my idea), but high-level theories, such as special relativity, do not disappear overnight in the face of the first apparently adverse result. One must account for this persistence, for as a matter of fact almost all theories live to a greater or lesser extent with some degree of anomaly unresolved in their comparison with experiment. Lakatos' idea is to define a research programme. This programme has a central hard core of ideas which are non-negotiable as far as that programme is concerned. They define what it is about. In the Newtonian account of the solar system, the hard core was universal inverse-square law gravity. In the comparison with experiment and observation, the hard core is buffered by a 'protective belt' of auxiliary hypotheses which can be adjusted as necessary to improve the

[79] cf. Thomson (1987).
[80] See Polkinghorne (1989b), ch. 21.
[81] Polanyi (1958), p. vii.
[82] Lakatos (1978), pp. 8–101; for a useful critique, see Newton-Smith (1981), ch. 4.
[83] Popper (1959); (1963).

empirical fit. When Uranus is found not to be behaving according to prediction, we do not abandon Newtonian gravity, but we introduce the hypothesis of an outer, unobserved, planet whose influence is perturbing the motion of Uranus. The subsequent discovery of Neptune represents the kind of 'stunning, dramatic and unexpected' discovery which Lakatos takes as the endorsement of a research programme's being worthy of the cachet 'progressive'. (He later extended this to include the unforced retrospective explanation of known but hitherto puzzling phenomena.) However, research programmes can become degenerating, as the cost in auxiliary hypotheses exceeds the gain of preserving the hard core. When discrepancies were discovered in the motion of Mercury, explanation by invoking an innermost planet, Vulcan, unobserved because of its closeness to the Sun, ran into more and more difficulty. After two centuries of great success, the Newtonian research programme began to go downhill. Eventually it was replaced by Einstein's general relativistic programme, which not only explained in an uncontrived manner the anomaly in the perihelion of Mercury, but also scored a 'stunning, dramatic and unexpected' success in predicting accurately the deviation of starlight by the Sun's gravitational field.

Clearly, this is a more persuasive mould into which to fit scientific history than a simple Popperian story of bold conjecture and unchallengeable refutation. Equally clearly, it has paradigmatic appeal to the theologian, for it endorses a certain fluidity of explanation on detailed points (of which, for example, prophetic history and theodicy can both take advantage) while at the same time granting respectability to a tenacious holding of central truth (the theological hard core of the existence of God and his self-manifestation in Christ). Theology is a scientific research programme!

Perhaps the most enthusiastic and extensive use of Lakatosian ideas in theology has been attempted recently by Nancey Murphy.[84] The difficulty will always be to substantiate the claim to the generation of stunning new results. Murphy uses Catholic Modernism as a test case, characterizing it as 'progressive' on the basis of George Tyrrell's 'predictions' that scientific history would not contradict biblical history on fundamental points, and that papal absolutism would prove a passing phase. The case looks a little thin.

There are two contrasting difficulties with the Lakatosian approach. One is that it has a flexibility which will enable it to fit a great many too many instances. I have a core belief that Arsenal is the best football club in the country. I protect this belief in the face of a run of failures by appeal to the auxiliary hypotheses of referee's incompetence, bad luck with injuries, and so on. Because they are in fact a decent side, occasionally they win the FA Cup or top the league, and then I score a stunning success.

[84] Murphy (1990).

The other difficulty is the reverse one, that in fact Lakatos' idea does not fit all actual scientific research programmes. In particular, historico-scientific theories much more frequently convince by casting new light on an extensive body of old facts than by making stunning new individual discoveries (though I must admit that George Gamow's primitive big-bang cosmology did predict a kind of cosmic background radiation, though that was largely forgotten till Penzias and Wilson made their discovery). A similar point is also made by Michael Banner in relation to Darwin's theory of evolution.[85] It is a common suggestion that it is these broad explanatory theories in science, which convince by their inference to the best explanation of general observations, rather than by the prediction of novel individual effects, which bear the closest kinship to theology. I conclude, therefore, that the Lakatosian analogy is probably only of limited value to theologians.

Finally, we should note that many people detect that a change in philosophical climate has been taking place in recent years. We are said to have entered the era of 'post-modernism'. Like most slogans, the phrase has a somewhat plastic character which enables people to put it to a variety of uses. I think that Nancey Murphy has given a useful characterization of what essentially is involved.[86] Three strands are woven together to form the post-modernist thread. Knowledge is conceived in the form of a holistic world-view rather than as arising from the discovery of unquestionable foundations on which to erect its edifice. Language is to be understood in terms of usage (Wittgenstein) rather than in terms of reference and the representation of reality. There is an irreducible role for the community, which is more than an aggregate of individuals. Post-modernism is a philosophy of the package-deal. It shows its influence in the theology of George Lindbeck[87] and the ethical theory of Alasdair MacIntyre.[88]

Clearly there are gains in this shift to the whole. Theology, with its emphasis on the role of the reception of doctrine by the whole Church and the role of worship in forming its liturgy-assisted logic,[89] is bound to recognize that. Yet there are also dangers of loss at the same time. Post-modernists tend to prescind from questions of truth, as we can see from the coyness exhibited by Lindbeck about the truth claims of theology. Typical of post-modernism is the revived pragmatism of Richard Rorty,[90] which in

[85] Banner (1990), ch. 6.
[86] Murphy (1990), pp. 199–208.
[87] Lindbeck (1984); for a critique, see Polkinghorne (1991), pp. 12–14.
[88] MacIntyre (1981).
[89] Polkinghorne (1991), p. 18.
[90] Rorty (1980).

William James's phrase sees 'the true is the good in the way of belief'[91] and rapidly leads on to the situation in which 'The pragmatist does not wish to explicate "true" at all, and sees no point in the absolute-relative distinction, or in the question of whether questions of appraisal *genuinely* arise.'[92] I think science and theology can make common cause in opposing decline into a merely intellectual utilitarianism and in insisting on the pursuit of the difficult but essential task of seeking to understand what *is*.

Religious belief is not merely a disguised way of expressing motivation for conduct. But neither is it merely intellectual assent to propositions about reality. It involves both 'believing that' and 'believing in', inextricably intertwined. The Christian creeds arose as baptismal *symbola*, confessions at the threshold of commitment. They call for the obedience of the will as well as the recognition of what is the case.

A NOTE ON SCIENTIFIC REALISM VERSUS EMPIRICAL ADEQUACY

In my defence of scientific realism, based on an analysis of a sustained episode in the history of elementary particle physics,[93] I appealed, as many others have done, to the notion of abduction or inference to the best explanation. Bas van Fraassen has asserted that science can aspire to no more than empirical adequacy, rather than verisimilitude, and its theories can command no more than acceptance, rather than belief.[94] An essential step in his discussion is to deny the validity of inference to the best explanation. A key argument for this is that abduction 'is a rule that only selects the best among the historically given hypotheses. We can watch no contest of the theories we have so painfully struggled to formulate, with those no one has proposed. So our selection may well be the best of a bad lot.'[95] The truth may lie hidden in the submerged ideas we are incapable of forming.

Two replies are in order. One is that implicit in this criticism is the notion, common to philosophers, that there are really lots of 'good' theories (i.e., economic, fruitful, in accord with general principles, etc.) which are capable of saving the phenomena. The experience of scientists of how hard it is to find one such theory, does not really encourage this point of view. Of course, that difficulty might be due to our intellectual blinkers, but it is striking that within the rational limitations we find natural to impose, we are nevertheless led time and again to an intellectually satisfying, and

[91] Quoted in McMullin (1988), p. 66.
[92] ibid., p. 68.
[93] Polkinghorne (1989b), ch. 21.
[94] van Fraassen (1980); (1989).
[95] van Fraassen (1989), p. 143.

essentially unique, fundamental understanding. As an empirical fact, abduction actually appears to work. We succeed in finding what certainly appear to be best explanations.

That observation leads on to the second point. Van Fraassen acknowledges that there could be a response to his objection based on human privilege: 'Its idea is to glory in the belief that we are by nature predisposed to hit on the right range of hypotheses.'[96] This would depend on the medieval principle of *adequatio mentis et rei*, the congruence of the mind with the way things are. Van Fraassen rightly rejects a justification of this principle on evolutionary grounds (our survival cannot depend on a capacity rightly to frame a theory of quarks). He mentions Plantinga's claim that this *adequatio* could be defended by a belief that we are made in God's image, but rather feebly rejects this because he thinks God would not particularly want us to understand deeply his physical creation. Why not?

I think that the history of science suggests most strongly that we can successfully infer to the best explanation and thereby gain verisimilitudinous knowledge of physical reality. We experience the universe's rational transparency to us, and this wonderful intelligibility is indeed made comprehensible by the insights of natural theology.[97] One sees once again that there is a supportive interaction between scientific and religious understanding.

[96] ibid.
[97] Polkinghorne (1988), ch. 2.

3

Divinity

'One God the Father Almighty'

The question of the existence of God is the single most important question we face about the nature of reality. Anthony Kenny says:

> After all, if there is no God, then God is incalculably the greatest single creation of the human imagination. No other creation of the imagination has been so fertile of ideas, so great an inspiration to philosophy, to literature, to painting, sculpture, architecture, and drama; no other creation of the imagination has done so much to stir human beings to deeds of horror and nobility, or set them to lives of austerity or endeavour.[1]

Is that what God is, a gorgeously fertile figment of the imagination, with man the true reality (as Feuerbach believed), or has the thought of him been the inspiration of so much human creativity precisely because he is the creative ground of all that is?

I like the story of a radical English theologian who had been giving a lecture to a group of clergy. At the end, one of them said, 'Professor X, do you believe in God?' He received a carefully nuanced academic reply. 'No, no,' said his interrogator, 'I just want to know if you believe in God.' Professor X then said, 'I believe, indeed I know, that at the heart of reality is One who reigns and loves and forgives.' It was a splendid reply, going with unadorned directness to the heart of what belief in the existence of God is all about. David Pailin says that 'a theistically satisfactory concept of the divine must conceive of the divine as being intrinsically holy, ultimate, personal, and agential'.[2] Does such a concept of God make sense? If so, do we have reason for believing in such a being?

Neither question is easy to answer. God is a different kind of being from any other that we might speak about. He is not part of that metaphysical

[1] Kenny (1989), p. 121.
[2] Pailin (1989), p. 24.

monism we considered in chapter 1, for his active will is the sustaining ground of that single created reality. Diogenes Allen says, 'God is not the final member of a succession of beings studied in cosmology or in any of the sciences, any more than God was the top storey of Aristotle's hierarchical universe, the unmoved mover.'[3] We have to use the language of finite being to try to talk about the infinite – we have no other means at our disposal – but it will always have to be in some stretched, analogical sense. It is less misleading to speak of God as personal rather than impersonal (he is more like a person than a 'force'), but that is not licence for the *naïveté* of the picture of an old-man-in-the-sky.

If the notion of God were incoherent, he could hardly exist. But I am cautious about our powers to assess coherence, particularly when we move into realms of experience far removed from the preserve of everyday thought. In 1900, any competent first-year philosophy undergraduate could have demonstrated the 'incoherence' of anything appearing sometimes like a wave and sometimes like a particle. Yet that is how light was found to behave and, through the universe-assisted logic of that discovery, we have been led to the invention of quantum field theory, which combines wave-like and particle-like behaviour without a taint of paradox.[4] Common sense (and a good deal of philosophy, particularly in the twentieth century, is a kind of painstaking common sense) is not the measure of everything. Thus, though the first, the conceptual, question has logical priority, I attach greater importance to the second, the evidential, question of whether we have reason to believe God actually exists. These lectures – indeed the whole sequence of Gifford Lectures stretching back to 1888 – are attempts to address that issue. The paradox lies in the fact that He who is most real is also He who is most elusive. Usually we can gain some purchase on the question of the existence of entities by comparing instances of their claimed presence with instances of their acknowledged absence – or, if like gravity, the entity is always present, at least instances of variation in the strength of its effects. But God is always present and he neither waxes nor wanes. The One who is the ground of all that is must be compatible with all that is. The nearest analogy in the physical world would be a universal medium, such as the nineteenth-century aether or the twentieth-century quantum vacuum. The former faded away when relativity theory left it no work to do; the latter provides a basis for understanding certain pervasive effects (vacuum polarization) and even (if extreme cosmological speculation be true) the origin of this particular universe that we experience (though the vacuum itself remains unexplained). The atheist will think God is like the aether. I

[3] Allen (1989), pp. 74–5.
[4] See Polkinghorne (1979), ch. 5.

think the better analogy is with the quantum vacuum. We have already referred to natural theology's appeal to a divine explanation of the intelligibility and anthropic 'fine tuning' of the universe. Further general considerations lie ahead of us. And if God is personal, then his presence, though unfailing, will not have the dreary uniformity of the action of a force, but it will manifest itself in ways that are appropriate to the individuality of circumstance. Hence we shall investigate the testimony of revelation, understood, in a way which I hope Lord Gifford would be able to accept, as reference to events or people particularly transparent to the divine presence, rather than understood as some mysteriously endorsed knowledge of what otherwise would be ineffable.

Before we tackle that task, we should not scorn considering philosophy's aid to conceptual clarity. I do not deny its utility. I only protest against any claim to prejudge the issues of experience.

It is a difficult area for an amateur like myself to wander into. The technical discussion often attempts a hard clarity where one might think that more employment of chiaroscuro was called for. It might be useful to recall the remark of Niels Bohr that there are 'two sorts of truth: trivialities, where opposites are obviously absurd, and profound truths, recognized by the fact that the opposite is also a profound truth'.[5] In the latter case, we must strive for coherence by seeking to delimit the range of applicability of the contrasting concepts (as in the case with complementary notions in quantum theory, from which Bohr drew his insight).

It is part of the classical Christian tradition, stemming particularly from St Thomas Aquinas, to lay stress on the simplicity of the divine nature. Naturally this does not mean a facile rational transparency – Pailin rightly says that 'a "simple" theistic faith is a contradiction in terms. God is not that kind of an object'[6] – but an unanalysable unity of being. David Burrell interprets this unity as required, because if there were divine parts they might seem formal causes of the divine nature, which is contradictory in a Being with aseity (being-in-itself).[7] Anyway, do not Christians, with Jews and Muslims, proclaim their faith in One God? Yet the main import of that proclamation is surely to assert that there is one prevailing will behind the world's existence and so to free us from the ambiguity of a dualism of light and darkness. It is not to make a metaphysical point about the divine nature. An unrelenting emphasis on divine simplicity easily translates into an image of static undifferentiated perfection, which is inhospitable to the use of personal language about God as Father. How does such an unchanging Unity

---

[5] Mackay (1977), p. 21.
[6] Pailin (1989), p. 41.
[7] Burrell (1986), p. 40.

relate to his changing creation? If we are forbidden to discriminate God's will from God's knowledge, how are we to understand those sinful acts, of whose occurrence he must have perfect knowledge but which are surely contrary to his perfect will? In our human experience we are aware that the unity of a person is not incompatible with the existence of parts within the psyche, as the insights of modern depth psychology, brilliantly anticipated by the introspective genius of St Augustine, make clear to us. The Christian doctrine of the divine Trinity is exciting, not only as doing justice to the varied economy of our experience of God, but also as the hint of how that property of structured-unity, encountered in ourselves, might extend into the ineffability of divine essence, neither confusing the Persons nor dividing the Substance. *Perichorēsis*, the mutual indwelling of the divine Persons which retains unity in diversity, might without impiety be considered under the metaphor of the 'divine bootstrap', the self-sustaining exchange constituting the aseity of the God whose quintessential nature is love. Something more than the banality of a scientist's metaphor is involved here. Although bootstrap ideas have not proved in the end to be successful in physics, the concept offers a possibility, unenvisaged by classical logicians, of declining to differentiate cause from effect, and so meeting the objection cited above from Burrell. Ward says of Thomist thought that

> Its basic error is in supposing that God is logically simple – simple not just in the sense that his being is indivisible, but in the much stronger sense that what is true of any part of God is true of the whole. It is quite coherent, however, to suppose that God, while indivisible, is internally complex.[8]

I have already spoken several times of God's aseity, that his essence implies his existence, so that his nature is that of a necessarily existing being, needing no explanation in terms of anything else. Concepts of divine necessity can function in a variety of ways. The most straightforward manner is as the ultimate answer to Leibniz's great question, 'Why is there something rather than nothing?' Every chain of explanation has to have a starting-point which is necessary to it, in the sense of its being the unexplained, without the assumption of which it would be impossible to frame an explanation at all. Intellectually, it is true that nothing comes of nothing. God can play that foundational role for a believer, but for the atheist it would be natural to follow Hume and take the existence of the physical world, with its intrinsic properties, as the ground of explanation. To do so, however, would not be to treat the physical world as necessary in quite the sense in which many theologians want to treat God as necessary.

[8] Ward (1982a), p. 216.

To approximate to the latter, one would have to claim that matter was somehow sufficiently self-explanatory. In fact the physical universe, by its very rational order and fruitfulness, seems to many to point beyond itself, so that there is more intellectual satisfaction in attributing its existence to the will of a self-sufficient agent than in treating it as a fundamental brute fact. There is a tendency today among some fundamental physicists to believe that there is a unique Theory of Everything (a TOE, as they lightheartedly say) whose discovery is just around the corner and which will then somehow explain why the world is. I very much doubt that ultimate explanation will prove attainable, both because the history of science discourages the expectation of such final achievement (in the past there has always been a surprise waiting around the next corner) and also because those who hope for a TOE have already tacitly decided that it must incorporate quantum theory and gravity, which are empirically necessary but not logically necessary requirements. Even if I am wrong about finality, a TOE will be more like the precise statement of Leibniz's question, rather than its answer. The most that physical theory could achieve (either in a modern TOE or in the old-style bootstrap programme) is a self-consistency and not a self-sufficiency. We would still want to ask, Why are there things which work in this particular way? In Stephen Hawking's words, 'Even if there is only one possible unified theory, it is just a set of rules and equations. What is it that breathes fire into the equations and makes a universe for them to describe?'[9] In the final paragraph of his book, Hawking seems to me to backtrack and suggest that possession of a TOE might somehow facilitate the discussion of why the universe exists.[10] I think that to suppose that is just a category mistake. Physics influences metaphysics in various ways, but it is not the same as metaphysics, and Leibniz's question is the ultimate *metaphysical* question. God's will can be its satisfying answer because the multivalued nature of the world's reality, in its order, beauty, ethical imperative and experience of worship, reflects the personal character of a Creator who is rational, joyful, good and holy. The plain assertion of the world's existence leaves unresolved the issue of how our diverse kinds of experience relate to each other. The strategy of materialist atheists is usually to claim that science is all, and that beauty and the rest are merely human constructs arising from the hard-wiring of our brains. I cannot accept so grotesquely impoverished a view of reality. Theism explains much more than a reductionist atheism can ever address. It is worth noting also that, as

[9] Hawking (1988), p. 174.
[10] ibid., p. 175. Steven Weinberg (1993) makes the interesting suggestion that a final theory, though not logically necessary, might be 'logically isolated' so that 'there is no way to modify it by a small amount without the theory leading to logical absurdities' (p. 189).

Talcott Parsons put it, in the history of humanity 'Religion is as much a human universal as language.'[11] Modern Western unbelief has something of the air of a cultural aberration in its rejection of a spiritual dimension to reality.

The strongest sense of divine necessity would be the claim that once one has logically understood the implication of the concept of God as 'one than whom no greater can be conceived' then one perceives that such a being must exist in any possible world. This, of course, was the breathtaking assertion of St Anselm in his ontological argument, given in the *Proslogion*. Debate on the validity of that argument has continued down the centuries. A scientist cannot but feel suspicious of attributing such power to unassisted reason. If we cannot even prove the consistency of arithmetic, it seems a bit much to hope that God's existence is easier to deal with. I do not doubt that Anselm's maximally perfect being, *if he exists*, will exist necessarily and will be in no ontological debt to anything else, but the question is whether such a being is actually instantiated. There is something deeply satisfying about the idea of God as the metaphysical keystone of the arch of being, but it is difficult to believe that to be a matter that can be settled by logical argument. Charles Hartshorne is a notable defender of the ontological argument, and he says, 'God is the *one individual conceivable a priori*,'[12] but that is not the same, it seems to me, as saying he is the one individual whose non-existence is inconceivable a priori. He is ontologically necessary, but not logically necessary. 'From the fact that one believes that God exists by necessity, it does not follow that one can establish that God exists simply by analysing the concept of God.' Equally, however, it may well be that 'the fact that we seem to be able to imagine God's non-existence merely shows that we are unable to imagine God'.[13] The words are Ward's. His belief is that the ontological argument depends upon the existence of value (to sustain the notion of a maximally perfect being) and that the coherence of the concept of God requires also appeal to the intelligibility of the world. Because these are contestable judgements, his conclusion is that 'The idea of God is the idea of a being which is either existent or impossible.'[14]

It is time to remind ourselves again of the tradition of apophatic theology, more consistently represented in Christian understanding by the thought of the Eastern Church than in the rational confidence of the Latin Church of the West. For apophatic theology the symbol of encounter with the divine is the clouds and darkness into which Moses entered on Mount Sinai. It

---

[11] Quoted in Hick (1989), p. 21.
[12] Hartshorne (1948), p. 31.
[13] Ward (1990), pp. 10–11.
[14] Ward (1982a), p. 33.

emphasizes the otherness, and so the unknowability, of God: 'He is not that.' Of course, we are never going to catch God in nets spread by our finite minds, but simply to acknowledge that would be to resign from the theological task altogether. It may well be, as Paul Tillich said, that 'God is symbol for God',[15] and to suppose otherwise is idolatry, but the Unknowable has acted to make himself known and the Christian believes that the perfect ikon of God in human terms is Jesus Christ. Charles Hartshorne protests against a merely negative theology, engaged in as a strategy somehow to cope with the necessary limits of human finitude. We should not make the mistake of 'the metaphysical false modesty of seeking to honour deity by refusing to apply any of our positive conceptions to him'.[16] A theologian must think, and negative concepts are as much human concepts as positive ones are.

Nevertheless, further perplexities arise when we assign qualities to God. To say God is good cannot be just a tautology, making God a kind of celestial dictator whose will is 'good' by mere definition, and woe betide those who question it. Yet if we mean that God is adjudged perfect on some independently existing scale of goodness, are we not making that goodness prior to him and so superior to him? It is a rather similar problem to that which we encountered when considering the simplicity of God. The answer must lie, I think, along similar lines, namely in a refusal to refrain from making differentiated statements about God but at the same time insisting on the self-sustaining integrity of the divine nature, so that God and his goodness are neither arbitrarily identical nor absolutely separable. I suspect this is what philosophical theologians are getting at in their celebrated equation of divine essence and divine existence – not just that the divine is Being with a capital B, but that God is self-subsistent perfection, identifying within himself not only cause and effect in the quality of aseity, but also supreme goodness and its instantiation in a divine bootstrap of virtue. Ward says of divine goodness that it is 'a necessary part of the being of God and could not exist as necessary in isolation from the totality of Divine nature'.[17] If all that is right, John Leslie's extreme axiarchism (the creative effectiveness of supreme ethical requiredness) is not a Neoplatonic 'Originating Principle' which might have as a consequence, in some emanating and descending chain of being, that there was 'an all-powerful person, an omniscient Designer',[18] but it is properly to be understood, purely and simply, as an insight into the divine nature itself.

[15] Quoted in Pailin (1989), p. 96.
[16] Hartshorne (1948), p. 35.
[17] Ward (1982a), p. 177.
[18] Leslie (1989), p. 166.

Similar thoughts will arise in connection with other divine qualities. David Burrell speaks of Maimonides' 'insistence that God is wise, but not "by wisdom". That is, God's *manner* of being wise is such that being God is the very norm and source of wisdom.'[19] The same would be true of divine knowledge of necessary truths, such as those of mathematics. God is not a rational tyrant who could have decreed that $2+2=5$, nor is he the perfectly rational, constrained improperly by the truths of reason, for they are not 'that *by which* God understands, but that *which* God understands'.[20]

Among the most puzzling, and the most pressing, of general questions about God are those concerned with how he is to be understood to relate to time.[21] It is clear that there must be an eternal pole to the divine nature. His steadfast love cannot be subject to fluctuation if he is worthy of being called divine. Emphasis on this alone would lead us to a static picture of God, but could that be true if the nature of love is relatedness and that to which God relates, namely his creation, is itself subject to radical change? Unchanging divine benevolence will surely express itself differently in relation to a universe which is an expanding quark soup than in relation to a universe containing self-conscious sinful beings. The classical answer to that is to say that God relates to the whole of cosmic history 'at once'. The quark soup and sinful humanity are equally present to the One who, in Aquinas's phrase, does not foreknow the future but, from the perspective of his eternity, simply knows it. Augustine said, 'God does not see things piecemeal, turning his gaze from side to side; he sees everything at once.'[22]

A certain concept of the nature of perfection underlies this point of view. Part of the agenda for resistance to a more positive engagement of God with time is because that would seem to imply change, and how could that occur without jeopardizing the divine excellence? A maximal being has nowhere else to go, as he sits perched on the topmost pinnacle of metaphysical eminence. But, just as in physics we can conceive of an equilibrium which is not simply a static staying put but which is the dynamical exploration of a pattern of possibility (atoms are something like that), so we can surely conceive a dynamical understanding of perfection, which resides, not in the absence of change, but in perfect appropriateness in relation to each successive moment. It is the perfection of music rather than the perfection of a statue. I do not think that God is necessarily simply eternal, so that he can only relate to time in a holistic way.

Yet it could be argued that the single divine view of cosmic history is

[19] Burrell (1986), p. 53.
[20] ibid., p. 67.
[21] Polkinghorne (1989a), ch. 7.
[22] Quoted in Kenny (1979), p. 25.

necessary if God is to be in the right relationship to his creation. Only if he relates to it in its temporal totality can he fully exercise that providential care which is appropriate to him.[23] Only if in his eternity he knows simultaneously that tomorrow I shall pray for a particular outcome and that today my friend is making a decision relevant to that outcome, can he really be a God capable of responding to prayer in influencing that decision. It is claimed additionally that, because God knows both of those decisions *as they happen*, they are both still decisions of free agents. A God incapable of exercising a timeless free response to a multitude of temporal free actions is just a God condemned to react to things as they happen, doing the best he can but continually having to revise his plans in the light of changing circumstances. Only a God who sees all that was, and is, and is to come, 'at once', is able to produce the best for his creation. That is how the argument goes.

There are two things I would want to say in reply. The first is to ask whether the universe has the appearance of being maximally planned in that timeless way, or whether it has more the appearance of an unfolding process within which God is certainly at work but, in Arthur Peacocke's striking phrase, as 'an Improviser of unsurpassed ingenuity'[24] rather than as the composer of a fixed cosmic score. David Hume remarked that 'Some small touches given to Caligula's brain in his infancy might have converted him into a Trajan,'[25] and one wonders why a God who *knows* what Hitler will do did not take that into account in some way. I shall address the problems of theodicy in the next chapter, but it seems to me that they are greatly intensified by the notion of timeless divine action.

The second difficulty is the question of whether an eternal view of temporal history, all at once, is a coherent idea. It was certainly the picture adopted by classical physicists when they drew spacetime diagrams of physical processes. It was as if they were looking down on spacetime from some external and eternal dimension, as the God of classical theism might be supposed to do. But that God's-eye-view was possible for them because they were describing a *deterministic* world, in which there is in fact no real distinction between past, present, or future (as Einstein indeed proclaimed in a letter written to a friend, just before his death). If we know the present in such a Laplacean world, we can with perfect accuracy predict the future or retrodict the past. Time is a kind of parameter marking where we are along the temporal tramline, but with no real significance, since there is no real change, only rearrangement. Such a world is a world of static being,

[23] I am indebted to Peter Burrows sj for a helpful conversation on this point.
[24] Peacocke (1986), p. 98.
[25] D. Hume, *Dialogues Concerning Natural Religion*.

without true becoming. That is not the world that we human beings live in, nor is it the world described by modern science. (I have argued that case extensively elsewhere.[26]) This puts in question the coherence of an eternal view of such a world.[27]

The future is not up there waiting for us to arrive. We make it as we go along. H. P. Owen was surely right to say that 'Temporal events cannot be known timelessly if they are to be known as they really are.'[28] The trouble with the timeless view of God's relationship with his creation is that it is in danger of denying the reality of becoming: it 'implies that what appear to be the *processes* of reality are in fact an illusion. In reality – that is, in the ultimate state of reality as known totally by God – all events are simultaneous.'[29] It seems to me that it is not sufficient for God just to know that events are temporal, he must know them as temporal in their due succession. David Pailin goes on to claim that such denial of the reality of process leads to a denial of the reality of a personal relationship, for it makes God's knowledge of us like our (partial) knowledge of historical characters, rather than our developing knowledge of our friends. Willem Drees raises the question how God could enjoy music without a genuine experience of temporal sequence.[30]

There is also a philosophical difficulty in relation to how a necessary and timeless God could be the Creator of a contingent and temporal world. Ward says, 'the truly contingent cannot arise from the wholly necessary',[31] and his conclusion is that 'If genuinely free creatures are admitted, there is an overwhelmingly strong argument against Divine immutability and for Divine temporality.'[32] We return to the notion of internal complexity within divine indivisibility. 'The general solution is to say that there is just one individual, possessing both a set of necessary properties and a set of contingent properties.'[33]

Thus, I am persuaded that in addition to God's eternal nature we shall have to take seriously that he has a relation to time which makes him immanent within it, as well as eternally transcendent of it.

The view of God's eternal grasp of the whole of time, which I have been criticizing, is part of what is often called classical theism. Its major concern is

[26] Polkinghorne (1988), ch. 3; (1989a), chs 1 and 2; (1991), ch. 3.
[27] See also the discussion of 'The Block Universe' by C. J. Isham and J. C. Polkinghorne in Russell et al. (1993).
[28] Quoted in Fiddes (1988), p. 91.
[29] Pailin (1989), p. 85.
[30] Drees (1990), p. 150.
[31] Ward (1982a), p. 3.
[32] ibid., p. 51.
[33] ibid., p. 165.

to preserve deity inviolate from the influence of the transient. Divine impassibility (which is correctly concerned to safeguard our idea of God from any magical taint that suggests he can be manipulated from outside himself) is translated into a chilling isolation from his creation. Thus Aquinas can say that 'Being related to God is a reality in creatures, but being related to creatures is not a reality in God.'[34] This seems to purchase divine freedom at the price of divine indifference. Its God seems more like the God of Stoicism than the God of Christianity. Hartshorne says of 'the assumption that the best or optimal dependence is zero dependence' that 'I dare affirm that this assumption is a blunder so great – and so influential – that one might scarcely hope in the whole history of thought to find a greater or more influential.'[35] It certainly accords ill with the God of the Bible: 'How can I give you up, O Ephraim! How can I hand you over, O Israel!' (Hos. 11.8).

Once God is acknowledged to be vulnerable through his love for his creation, it becomes possible to speak of the mystery of a suffering God. Paul Fiddes says that 'We must try to think of a God who can be the greatest sufferer of all and yet still be God.'[36] There is a tension between speaking of God both as fellow-sufferer and as the ground of a sure hope. 'If it is essential that a God who helps us should sympathize with our suffering, it is also essential that he should not be overcome or defeated by suffering.'[37] His vulnerability has been freely embraced. It arises from the fact that his creation matters to him, not because it has intrinsic power over its Creator, but (in the beautiful metaphor of W. H. Vanstone[38]) in the way that an adopted child has become indispensable to the family into which he or she is freely accepted. It is this divine acceptance which is the ground of the hope of the redemption of suffering: 'out of the desire for his creatures he chooses to suffer, and because he chooses to suffer he is not ruled by suffering; it has no power to overwhelm him because he has made the alien thing his own'.[39] Yet this does not mean that divine suffering is in any way more protected and less real than human suffering. Fiddes goes so far as to say, 'God's suffering must be not only a feeling and an act, but also an injury and constraint upon him.'[40] For the Christian, what this could mean finds its focus in the passion of Christ, understood as God's opening his arms to embrace the bitterness of the world. The purpose of Fiddes' book on the

---

[34] Quoted in Ward (1990), p. 21.
[35] Hartshorne (1948), p. 50.
[36] Fiddes (1988), p. 2.
[37] ibid., p. 100.
[38] Vanstone (1977), p. 69.
[39] Fiddes (1988), pp. 108–9.
[40] ibid., p. 262.

creative suffering of God is said to be to speak of 'a God who suffers eminently and yet is God, and a God who suffers universally and yet is present uniquely and decisively in the sufferings of Christ'.[41] The dialectic of a suffering God finds its historical expression in a crucified Messiah.

David Brown says of the concept of a suffering God:

> one must beware of too easy an anthropomorphism. For any suffering God endures will still be subject to the enormous transformation an omniscient consciousness brings. So much of human misery is due to uncertainty about the outcome or duration of the pain, from both of which God would be exempt.[42]

If we believe that God has a genuine temporal pole to his being, that qualification will be modified. Just as Jesus may be believed to have trusted in ultimate divine vindication but yet to have experienced truly the dereliction of Calvary, so God's engagement with suffering takes place in the context of the steadfastness of the divine will but not in the certainty of an immutable future laid out before him. One sees here something of a moral argument for divine temporality.

The opposite extreme to the divine detachment of classical theism would be pantheism, a thorough-going monism which identifies God with the world – *deus sive natura* (God or nature) in Spinoza's famous phrase – so that he becomes a kind of cipher for the marvellous rational order of the universe. This has been quite a popular religious stance among scientists. Einstein wrote that 'The scientist's religious feeling takes the form of a rapturous amazement at the harmony of natural law, which reveals an intelligence of such superiority that compared with it, all systematic thinking and acting of human beings is an utterly insignificant reflection.'[43] God seems to play a similar role in Stephen Hawking's *Brief History of Time*, popping up in the text from time to time, though absent from the index. Biologists may look in this manner at the fruitfulness of evolutionary history. Ralph Burhoe, for example, says that 'Man's salvation comes in bowing down before the majesty and glory of the magnificent program of evolving life in which we live and move and have our being.'[44]

I do not want to deny the genuine religious intuitions expressed in such sentiments, but they seem to me to be sadly inadequate as an account of the human meeting with the divine. I think that encounter to be characterized by an unmistakable otherness in One who stands over against us in judgement and mercy. 'Woe is me! For I am lost; for I am a man of unclean

---

[41] ibid., p. 3.
[42] Brown (1987), p. 45.
[43] Einstein (1948).
[44] R. Burhoe, *Zygon*, 10, p. 367.

lips . . . for my eyes have seen the King, the Lord of hosts!' (Isa. 6.5). If classical theism makes God too remote, pantheism makes him too domesticated.

Many people today believe the answer to lie in the half-way house of panentheism: the world is part of God, but he exceeds the world. Immanence and transcendence, closeness and the beyond, are said to be fused in this way. Hartshorne, after having defined panentheism to mean that God is 'both His system [the cosmos] and something independent of it', goes on to claim that 'Panentheism agrees with traditional theism on the important point that the divine individuality, that without which God would not be God, must be logically independent, that is, must not involve any particular world,' while also agreeing 'with the equally necessary point of traditional pantheism that God cannot in his full actuality be less or other than literally all-inclusive'.[45]

While I am sympathetic to what panentheism sets out to achieve by way of balance between divine transcendence and divine immanence, I cannot myself see that it succeeds in doing so in an acceptable way. It seems to me that the Christian doctrine of creation is rightly concerned with the self-surrender of divine all-inclusiveness in the creating of a world genuinely other, to which God can be 'closer than breathing', in the sense of continuously being aware of it and interacting with it, without being, even partially, identified with it; that this is necessary in order to accommodate that religious experience of being separate from the divine to which I have referred; and that without it the problems of evil and suffering, and of the scientifically probable finiteness of past cosmic history, become greatly exacerbated. As I have tried to explain elsewhere,[46] I find the theology of Jürgen Moltmann, with its concept of God's 'making way' for something other than himself, to be extremely helpful.[47] As I shall try to explain in a later chapter, I see panentheism as the eschatological destiny of creation, not its present status. One of the reasons for the popularity of the idea of panentheism in much current writing about science and religion is because it seems to offer an acceptable way of conceiving how God acts in the world. When that is thought through it often leads, I believe, to the notion of divine embodiment, that the world is in God but he exceeds it, in a way analogous to the way our bodies are part of us but we are more than our physical integument. A careful discussion of this idea has been given by Grace Jantzen,[48] but I have elsewhere criticized it as leading to one of two

[45] Hartshorne (1948), p. 90.
[46] Polkinghorne (1988), ch. 4.
[47] Moltmann (1981), pp. 105–14; (1985), pp. 86–93.
[48] Jantzen (1984).

unacceptable consequences: that God is too greatly in thrall to the changes
of cosmic history, or the universe is too greatly in thrall to divine
manipulation.[49] Often, however, the concept is kept in too soft a focus for
the difficulties to be readily apparent.[50]

A different realization of panentheism is provided by process theology,
inspired by the metaphysical ideas of A. N. Whitehead (see chapter 1).
Inclusiveness is achieved by the fact that the divine is party to every event, as
the reservoir of past experience, the presenter of present possibility and the
lure towards a particular future outcome. Yet initiative, in the form of
concrescence, lies with the event itself, so that a process-panentheism
expresses some distinctions which might be blurred by embodiment, though
at the cost of what appears to be a diminishment of the divine role.

There are, as we have already seen, considerable difficulties with
Whitehead's metaphysical scheme. Charles Hartshorne, the *doyen* of process
theologians, has expressed his reservations (he is prepared to use a suitably
nuanced language of embodiment), and David Pailin, perhaps the most
fluent British writer on the theme, has abandoned many aspects of
Whitehead's thought. He finds the notion of a divine envisagement of a total
portfolio of possibilities to be 'fantastic and incredible',[51] though his ground
– that there are an awful lot of such possibilities – sounds like an
unwillingness to take divine infinity seriously. He will allow God to be a
good club player at the game of cosmic chess, but not the supreme Grand
Master.

My own difficulties with process thought lie elsewhere. I think there are
very valuable insights contained within it, particularly in its emphasis on the
dipolar nature of God and his consequent true engagement with time. In
fact, I am quite often told that I am a process theologian, *malgré moi*. The
reason I decline the honour offered is that I do not find the God of process
theology to be an adequate ground of hope, and I believe hope to be central
to an understanding of what is involved in a Christian view of God's reality.
Whitehead spoke of God as 'the great companion – the fellow-sufferer who
understands'.[52] It is a noble vision, but it comprehends only part of the
Christian understanding, for the latter wishes also to speak of a time when
'God himself will be with them; he will wipe away every tear from their eyes,
and death shall be no more, neither shall there be mourning nor crying nor
pain any more, for the former things have passed away' (Rev. 21.3–4). Pailin
says that 'the proper goal of divine creativity is not to be envisaged as the

49 Polkinghorne (1989a), pp. 18–23.
50 Peacocke (1979): but the later discussion of Peacocke (1990) is in sharper focus and
seems to follow a line similar to that adopted here.
51 Pailin (1989), p. 61.
52 Whitehead (1978), p. 351.

attainment of a particular state of affairs but as the continual pursuit of aesthetic enrichment'.[53] For the process God, it seems, it is indeed better to travel hopefully than to arrive.

Such a view goes against the grain of religious experience. If I were asked to say what is the present consequence for the believer of holding a belief in God – what is its 'cash value' in terms of life now – I would summarize the answer by saying that there is One who is worthy of worship and that he is the fitting ground of the hope that there is a meaning to existence and a final fulfilment awaiting us.[54] The fundamental character of that intuition of hope is endorsed by many theologians. Barth said of the righteousness of God that it is 'the meaning of *all* religion, the answer to *every* human hope and desire and striving and waiting, and it is especially the answer to *all* that human activity which is concentrated upon hope'.[55] Even an extreme radical of the 'God-is-dead' school, William Hamilton, could write, 'To be a Christian today is to stand, somehow, as a man without God, but with hope,'[56] though I believe that hope is in fact dependent upon the belief that Hamilton discarded. Jürgen Moltmann wrote that 'from first to last, and not merely in the epilogue, Christianity is eschatology, is hope'.[57] He and Wolfhart Pannenberg – the 'theologians of hope', as they are sometimes called – emphasize that reality is proleptic, for the meaning of the present depends upon the future and, in Pannenberg's words, 'According to the Biblical understanding, the essence of things will be decided only in the future. What they are is decided by what they will become.'[58] Commenting on the theology of hope, William Taylor writes that 'The future is not a matter of some far-off divine event. The future is the reality that encounters us as we put our hope and anticipation to the test.'[59]

I have already explained that I do not believe that the future is up there waiting for us to arrive. If the future plays so significant a role in the present, it is not because we are witnessing the unfolding of an inexorable plan, but rather because final fulfilment, though arrived at through the contingencies of history, is guaranteed by the steadfast love of a God ceaselessly at work within that history, whose benevolent intention will not ultimately be thwarted.[60]

My dissatisfaction with process theology is due to the fact that I do not

[53] Pailin (1989), p. 132.
[54] cf. the particular experiences described and analysed in Taylor (1992), ch. 2.
[55] Green (1989), p. 129.
[56] Quoted in Gilkey (1969), p. 116.
[57] Moltmann (1967), p. 16.
[58] Pannenberg (1968), p. 169.
[59] Taylor (n.d.), p. 132.
[60] Polkinghorne (1989a), ch. 9.

think it presents a picture of God's action which is sufficiently strong to make him the ground of a reliable hope. Ian Barbour, who is notably sympathetic to process ideas, admits as much when he says, 'Process theology does call in question the traditional expectation of an *absolute victory over evil*.'[61]

Pailin assures us that 'panentheism understands God to be eminently active as well as eminently passive'.[62] Hartshorne says, 'Religious faith imputes to God at least the kind and degree of power that the world needs as its supreme ordering influence'; but what that means is immediately qualified when he goes on to say, 'It comes to the same thing to say that divine power must suffice to enable God to maintain for himself a suitable field of social relations.'[63] He must be in touch ('the great companion'), but is that really enough?

Pailin tells us that 'God's activity is accordingly to be conceived as the luring influence of a love which respects the proper integrity and intrinsic value of others.'[64] Hartshorne believes that God is so important to us that his influence will be so great that it 'sets *narrow* limits to our freedom', and thus

> God can rule the world and order it, setting optimal limits for our free action, by presenting himself as essential object, so characterized as to weight the possibilities of response in the desired respect. This divine method of world control is called 'persuasion' by Whitehead and it is one of the greatest of all metaphysical discoveries, largely to be credited to Whitehead himself.[65]

Barbour says that 'process thinkers may sometimes seem to make God powerless, but in fact they are pointing to alternative forms of power in both God and human life'.[66]

I would not question that the old-style masculine, monarchical picture of God's rule – the cosmic tyrant against whom Whitehead rebelled – is in need of revision. In its place we are offered a more feminine mode, not itself wholly free from ambiguity (talk of persuasion can carry overtones of 'Do what you like, but you will break my heart if you do X'; the overbearing father is in danger of being replaced by the manipulative mother). But above all, it seems to me, we are being presented with a picture of God's activity which is congenial to those who believe the liberal fallacy that all one really

[61] Barbour (1990), p. 264.
[62] Pailin (1989), p. 89.
[63] Hartshorne (1948), p. 134.
[64] Pailin (1989), p. 94.
[65] Hartshorne (1948), p. 142.
[66] Barbour (1990), pp. 261–2.

needs in order to get things right is to present people with a clear picture of possibilities and then appeal to their better nature. Pailin tells us that 'God's saving activity may, for instance, be compared to the role of imaginative and courageous play-group leaders.'[67] But suppose the little blighters won't follow the bright suggestions offered, and instead of using the sand to build a model of the people's palace, they cram it down each other's throats? Whitehead says, 'God . . . does not create the world, he saves it; or, more accurately, he is the poet of the world, with tender patience leading it by his vision of truth, beauty and goodness.'[68] But, as the Fathers knew in their arguments with Marcion, if God is not Creator, he cannot be Redeemer either. The hope that process thought offers is summed up by Pailin in the words, 'The relation of God to history and humanity finds its best image as "a tender love that nothing be lost" (Whitehead).'[69] This is achieved by God's being the preserver of experience; the 'one achievement' is the 'enrichment of the divine life'.[70] We are offered the precise reversal of the splendid picture in the catechism, of humanity's destiny being to know God and enjoy him. The purpose of the universe seems simply to be the divine accumulation of realized and remembered possibilities. I do not find that an adequate account of hope.

Presumably God's consequent nature is as much the reservoir of the experiences of Hitler as of Mother Teresa. What are we to make of all the things we want lost (for future purgation is an important part of hope), of all the incompletenesses and wounds that we shall carry to our graves still needing their fulfilment and their healing? I do not want to be just a fly stuck in the amber of divine remembrance; I look forward to a destiny and a continuing life beyond death. To put it bluntly, the God of process theology does not seem to be the God who raised Jesus from the dead.

Accordingly, I feel I must grope for a more positive account of God's action than process thought affords. I have written about this search previously,[71] and I must again return to aspects of it in the next chapter. In brief, I believe that an important clue lies in making use of the conception of a role for active information-input in the open history of the world, which I sketched in chapter 1. This offers scope for positive divine interaction, hidden within the cloudiness of unpredictable process,[72] and fully respectful of 'the proper integrity and intrinsic value of others' (for I share

[67] Pailin (1989), p. 214.
[68] Whitehead (1978), p. 356.
[69] Pailin (1989), p. 164.
[70] Hartshorne (1948), p. 133.
[71] Polkinghorne (1989a); (1991), ch. 3.
[72] Ward (1982b, p. 95) argues for the theological necessity that divine action be indiscernible.

with process thinkers the recognition that this is a necessary component in any acceptable account of activity by Love). The veiled action of God within unpredictable process means that divine providence cannot be factored out from what is going on, with this set of events attributable to him and that set to natural causes. There is one web of occurrence in which all agencies interlace. I deny that this picture is an unacceptable return to a God of the gaps, for any gaps here are intrinsic to the nature of open process and are not mere, extrinsic, areas of ignorance. As such they are as much a constitutive part of our powers of free action as they are of God's.

Finally, let me say that it would not be enough to confine an account of God's activity to his relationship to humanity alone, even though that would seem to enable one to use the notion of persuasion without any appeal to a questionable panpsychism. David Bartholomew expresses the opinion in relation to divine action that he is 'disposed to think that the normal mode of [God's] action is in the realm of the mind'.[73] Three difficulties dissuade me from following that path. One, of course, is the general inadequacy of a merely persuasive, or even guiding, role for divine influence. I have already said enough about that. The second is that, if we accept a dual-aspect monism and regard human beings as psychosomatic unities, then there is no separate privileged realm of the mind to which God would have access in a way any less problematic than his relationship to the matter of his creation. And thirdly, such a view would suggest that the universe was on automatic pilot for fifteen billion years before the emergence of conscious beings. We seek an account of God where (in the stretched analogical language which theology has to use) he is rightly called Father (a symbol for steadfast and particular divine care, not interpreted in the sense of a narrow masculinity or patriarchy, but fully comprehending the maternal as well as the paternal) and rightly called Almighty (a symbol for divine action, not interpreted as the exercise of a cosmic tyranny, but consistent with God's loving respect for the freedom given to his creation). We need a concept of God's gracious agency which is powerful enough to sustain the hope of redemption from evil and gentle enough to correspond to the One whose service is perfect freedom.

One final word. Our rational inquiry must not blind us to the awe-inspiring and ineffable mystery of God to which religious experience bears testimony. Kierkegaard said, 'May we be preserved from the blasphemy of men who "without being terrified and afraid in the presence of God . . . without the trembling which is the first requirement of adoration . . . hope to have direct knowledge".'[74]

[73] Bartholomew (1984), p. 143.
[74] Quoted in Brown (1987), p. 17.

A NOTE ON AN ANALOGY FROM PHYSICS

For the theist, the rational beauty of the physical world is not just a brute fact, but a reflection of the mind of the Creator. Aesthetic experience and ethical intuitions are not just psychological or social constructs, but intimations of God's joy in creation and of his just will. Religious experience is not illusory human projection, but encounter with divine reality. There is an integrating wholeness in the theistic account which I find intellectually satisfying, even though it must wrestle with the mystery of infinite Being.

The theist and the atheist alike survey the same world of human experience, but offer incompatible interpretations of it. My claim would be that theism has a more profound and comprehensive understanding to offer than that afforded by atheism. Atheists are not stupid, but they explain less.

There is an interesting analogy with a case in physical science where theoretical physicists survey the same domain of physical experience but offer conflicting interpretations of it. Such instances are rare – despite the way philosophers of science talk about the underdetermination of theory by data – for it is seldom possible to find competing candidates in fundamental physics capable of offering ontologically incompatible accounts of equal empirical adequacy in relation to the same broad swathes of experience. Yet quantum theory provides an example of such a Wittgensteinian 'duck/rabbit': the 'Copenhagen' (indeterministic) and Bohmian (deterministic) accounts of quantum physics.[75] Both entail exactly the same (successful) empirical consequences, but they offer sharply contrasting ontologies of the quantum world. Despite there being no crucial experiment to discriminate between these two theories, almost all physicists adhere to the 'Copenhagen' point of view. It is the more mysterious (counterintuitive and unpicturable) account, but it is felt to be more intellectually satisfying because it offers an explanation of the Schrödinger equation which Bohm's theory must accept as brute fact. I believe that preference to be a rationally defensible one, and to bear some analogy with the theist's preference for his or her account of a wider reality. Although atheism might seem simpler conceptually, it treats beauty and morals and worship as some form of cultural or social brute facts, which accords ill with the seriousness with which these experiences touch us as persons.

[75] See Polkinghorne (1991), ch. 7, for an account.

4

Creation

'Maker of heaven and earth, of all that is,
seen and unseen'

In the beginning was the big bang. As the world sprang forth from the fuzzy singularity of its origin, first the spatial order formed, as quantum fluctuations ceased seriously to perturb gravity. Then space boiled, in the rapid expansion of the inflationary era, blowing the universe apart with incredible rapidity in the much less than 10^{-30} seconds that it lasted. The perfect symmetry of the original scheme of things was successively broken as the cooling brought about by expansion crystallized out the forces of nature as we know them today. For a while the universe was a hot soup of quarks and gluons and leptons, but by the time it was one ten-thousandth of a second old, this age of rapid transformations came to a close and the matter of the world took the familiar form of protons and neutrons and electrons. The whole cosmos was still hot enough to be the arena of nuclear reactions, and these continued until just beyond the cosmic age of three minutes. The gross nuclear structure of the universe was then left, as it remains today, at a quarter helium and three-quarters hydrogen. It was far too hot for atoms to form around these nuclei, and this would not occur for another half a million years or so. By then the universe had become cool enough for matter and radiation to separate. The world suddenly became transparent and a universal sea of radiation was left to continue cooling on its own until, fifteen billion years later, and by then at a temperature of 3°K, it would be detected by two radio astronomers working outside Princeton – a lingering echo of those far-off times.

Gravity is the dominant force in the next era of cosmic history. It continued its even-handed battle against the original expansive tendency of the big bang, stopping the universe from becoming too rapidly dilute but failing to bring about an implosive collapse. Although the early universe was almost uniform in its constitution, small fluctuations were present, producing sites at which there was excess matter. The effect of gravity enhanced these irregularities until, in a snowballing effect, the universe after a billion years or so, began to become lumpy and the galaxies and their stars began to form.

Within the stars nuclear reactions started up again, as the contractive force of gravity heated up the stellar cores beyond their ignition temperature. Hydrogen was burned to become helium, and when that fuel was exhausted a delicate chain of nuclear reactions started up, which generated further energy and the heavier elements up to iron. The elemental building-blocks of life were beginning to be made. Every atom of carbon in every living being was once inside a star, from whose dead ashes we have all arisen. After a life of ten billion years or so, stars began to die. Some were so constituted that they did so in the dramatic death-throes of a supernova explosion. Thus the elements they had made were liberated into the wider environment and at the same time the heavier elements beyond iron, inaccessible through the burning of stellar cores, were produced in reactions with the high-energy neutrinos blowing off the outer envelope of the exploding star.

As a second generation of stars and planets condensed, on at least one planet (and perhaps on many) the conditions of chemical composition, temperature and radiation were such that the next new development in cosmic history could take place. A billion years after conditions on Earth became favourable, through biochemical pathways still unknown to us, and utilizing the subtle flexible-stability with which the laws of atomic physics endow the chemistry of carbon, long chain molecules formed with the power of replicating themselves. They rapidly gobbled up the chemical food in the shallow waters of early Earth, and the three billion years of the history of life had begun. A genetic code was established, a biochemical alphabet in which the instructions for terrestrial life are universally spelled out. Primitive unicellular entities transformed the atmosphere of Earth from one containing carbon dioxide to one containing oxygen, thereby permitting important developments in metabolism. The process of photosynthesis evolved, *the* method by which the sun's energy is trapped and preserved for the maintenance of all living beings. Eventually, and then with increasing rapidity, life began to complexify through a process which certainly included the sifting of small variations through the environmental pressures of natural selection. Seven hundred million years ago, jellyfish and worms represented the most advanced forms of life. About three hundred and fifty million years ago, the great step was taken by which some life left the seas and moved on to dry land. Seventy million years ago, the dinosaurs suddenly disappeared, for reasons still a matter of debate, and the little mammals that had been scurrying around at their feet seized their evolutionary chance. Three and a half million years ago, the Australopithecines began to walk erect. Archaic forms of *homo sapiens* appeared a mere three hundred thousand years ago, and the modern form became established within the last forty thousand years. The universe had become aware of itself.

Such, in outline, is the story that science tells us about the history of the world. There are some speculations (particularly in the very early cosmology) and some ignorances (particularly in relation to the origin of life), but there seems to me to be every reason to take seriously the broad sweep of what we are told. Theological discourse on the doctrine of creation must be consonant with that account.

Of course, the first thing to say about that discourse is that theology is concerned with ontological origin and not with temporal beginning. The idea of creation has no special stake in a datable start to the universe. If Hawking is right, and quantum effects mean that the cosmos as we know it is like a kind of fuzzy spacetime egg, without a singular point at which it all began, that is scientifically very interesting, but theologically insignificant. When he poses the question, 'But if the universe is really completely self-contained, having no boundary, or edge, it would have neither beginning nor end: it would simply be. What place, then, for a creator?',[1] it would be theologically naive to give any answer other than: 'Every place – as the sustainer of the self-contained spacetime egg and as the ordainer of its quantum laws.' God is not a God of the edges, with a vested interest in boundaries. Creation is not something he did fifteen billion years ago, but it is something that he is doing now.

An important implication of the Christian doctrine of creation is that it clearly distinguishes the created order from its Creator. Barth says that 'Creation is the freely willed and executed positing of a reality distinct from God.'[2] Burrell says, 'What is at issue here is a clean discrimination of creation from emanation, of intentional activity from necessary bringing forth.'[3] Emanationism pictures the world as arising in a kind of panentheistic way, as the divine being's fruitfulness inevitably spills over into a multiplicity of consequences. In its view, the world is at the hem of deity.

Christian theology, on the contrary, sees the world as the consequence of a free act of divine decision and as separate from deity.[4] The universe's inherent contingency is conventionally and vividly expressed in the idea of creation *ex nihilo*. Nothing else existed (such as the brute matter and the forms of the classical Greek scheme of things) either to prompt or to

---

[1] Hawking (1988), p. 141.
[2] Green (1989), p. 188.
[3] Burrell (1986), p. 15.
[4] Pannenberg has even sought to ground the character of creation's independence in the self-differentiation of the Persons within the divine nature of the Trinity; see the discussion of Grenz (1990), pp. 85–7.

constrain the divine creative act. The divine will alone is the source of created being. 'In the doctrine of creation out of nothing, . . . Christians replaced the notion of irrational accident or blind chance by the concept of contingence.'[5] God's decision was freely made. This concept can be held to have played an important part in the ideological undergirding of modern science, for it implied both that the world was rational and also that the nature of its rationality depended on the choice of its Creator, so that one must look to see what actual form it had taken.

It is sometimes said that creation *ex nihilo* is just the sort of metaphysical speculation which got grafted on to biblical ideas when Christianity expanded into the late Hellenic world. It is certainly true that it is possible to give a natural exegesis of Genesis 1 which falls short of the explicit articulation of this concept, but I agree with Keith Ward that the doctrine is implicit in the clear claim that all depends upon God's will ('And God said, "Let there be . . ." '). 'It is therefore correct to see this doctrine of creation as implicit in the Biblical doctrine that God is the creator of heaven and earth, that he can do all things, that nothing is beyond his power.'[6]

The doctrine safeguards the fundamental theological intuition that creation is separate from its Creator, that he has made ontological room for something other than himself. Moltmann says, 'It is only God's withdrawal into himself which gives that *nihil* the space in which God becomes creatively active.'[7] On the other hand, Whitehead rejected the doctrine because he did not want God to play so absolute a role. 'He is not *before* all creation but *with* all creation.'[8] In their account of process theology, Cobb and Griffin tell us that it 'rejects the notion of *creatio ex nihilo*, if that means creation out of *absolute* nothingness . . . Process theology affirms instead a doctrine of creation out of chaos'[9] – which is certainly an exegetically possible view of what is involved in Genesis's reference to that which was 'without form and void' (*tohu wabohu* in Gen. 1.2). But once again I feel that process theology's diminished view of divine power does not allow God to be God.

Needless to say, lighthearted claims that modern physics has provided its own version of creation *ex nihilo* completely miss the point.[10] They are based on speculations about what might have happened in that intrinsically quantum cosmos before the formation of spatial order at the Planck time of

[5] Torrance (1989), p. 12.
[6] Ward (1990), p. 6.
[7] Moltmann (1981), p. 109.
[8] Whitehead (1978), p. 343.
[9] Cobb and Griffin (1976), p. 65.
[10] See, for example, Davies (1983), ch. 16.

10^{-43} seconds. We need to bear in mind the warning uttered by the great Russian theoretical physicist, Lev Landau, that his cosmologist friends were 'often in error but never in doubt'. All the same, bold speculators are sometimes right, and let us, for the sake of argument, suppose that they are correct in supposing that the universe of our experience has emerged, by one process or another, from a pre-existing quantum vacuum. Only by the greatest abuse of language could such an active and structured medium be called *nihil* (for in quantum theory, when there is 'nothing' there, it does not mean that nothing is happening[11]). It is just conceivable that physics may be able to show that given quantum mechanics and a certain gauge field theory of matter, then universes will appear; theology is concerned with the Giver of those laws which are the basis of any form of physical reality.

To hold a doctrine of creation *ex nihilo* is to hold that all that is depends, now and always, on the freely exercised will of God. It is certainly not to believe that God started things off by manipulating a curious kind of stuff called 'nothing'. There is no contradiction in holding at the same time a doctrine of *creatio continua*, which affirms a continuing creative interaction of God with the world he holds in being. The two are respectively the transcendent and the immanent poles of divine creativity. Peacocke says, 'The scientific perspective on the world and life as evolving has resuscitated the theme of *creatio continua* and consideration of the interplay of chance and law (necessity) led us to stress the open-ended character of this process of the emergence of new forms.'[12] That would not altogether have surprised St Augustine, who wrote that 'In the first instance, God made everything together without any moments of time intervening, but now He works within the course of time, by which we see the stars move from their rising to their setting.'[13] We do not today take so ready-made a view as Augustine expresses at the start of that passage – though elsewhere he suggested that 'In the beginning were created only germs or causes of the forms of life which afterward developed in gradual course.'[14] Of course, Augustine certainly believed that all was held in being by God's transcendent will: 'the universe will pass away in the twinkling of an eye if God withdraws His ruling hand'.[15]

The idea of *creatio continua* would not have surprised the prophet we call Second Isaiah, either: 'From this time forth I make you hear new things, hidden things which you have not known. They are created [*bārā*] now, not

[11] See, for example, Polkinghorne (1979), pp. 72–5.
[12] Peacocke (1979), p. 304.
[13] Quoted in McMullin (1985), p. 10.
[14] Quoted in Stannard (1982), p. 11.
[15] Quoted in McMullin (1985), p. 11.

long ago; before today you have never heard of them, lest you should say, "Behold, I knew them" ' (Isa. 48.6–7).

What consonance can we detect between these ideas and our scientific understanding? Clearly we cannot perform the ultimate experiment: remove the divine presence and see if the universe disappears. Belief in creation *ex nihilo* will always be a metaphysical belief, rooted in the theologically perceived necessity that God is the sole ground of all else that is. Belief in *creatio continua* can be more directly motivated by our perception of cosmic process, the evolving complexity of a universe endowed with anthropic potentiality. Freeman Dyson says, 'The more I examine the universe and the details of its architecture, the more evidence I find that the universe in some sense must have known we were coming.'[16] I cannot see what sense that could be other than the will of a Creator. I reject the strange claim of the Participatory Anthropic Principle,[17] that somehow observers bring about the grounds for their own existence, as being scientifically incredible. Attempts are sometimes made to appeal to quantum theory in defence of this idea. I think they fail because the influence of observers refers at most to the *outcomes* of quantum events and not to the basic laws governing those events, which have to have a certain structure in an anthropically fruitful universe – such as being quantum mechanical, for example.

Yet that fruitfulness is realized by the shuffling explorations of happenstance within the limits of possibilities constrained and preserved by natural law, which process we refer to in a slogan way as the interplay of chance and necessity. In 1860, the year after publishing *The Origin of Species*, Darwin wrote to his friend Asa Gray:

> I am inclined to look at everything as resulting from designed laws, with the details, whether good or bad, left to the working out of what we may call chance. Not that this notion *at all* satisfies me. I feel most deeply that the whole subject is too profound for the human intellect.[18]

Eight years later he was still writing to the same friend about the same problems. The redundancies and 'many injurious deviations of structure' did not look like the execution of the inexorable plan of an omniscient Creator. 'Thus we are brought face to face with a difficulty as insoluble as is that of free will and predestination.'[19] I think the comparison is apt. Part of a notion of *creatio continua* must surely be that an evolving universe is one which is theologically understood as being allowed, within divine providence, 'to make itself'. This seems to me preferable to saying that 'God the

[16] Dyson (1979), p. 256.
[17] Barrow and Tipler (1986), p. 22.
[18] Quoted in McMullin (1985), p. 139.
[19] ibid., p. 140.

Creator explores in creation',[20] for God cannot be thought to need to use the universe as an analogue computer to explore possibility. I have suggested that from a theological point of view the roles of chance and necessity should be seen as reflections of the twin gifts of freedom and reliability, bestowed on his creation by One who is both loving and faithful.[21] Does such anthropomorphic language make any analogical sense, applied to a universe which for virtually all its history has not contained conscious beings?

In chapter 1, I presented a picture of open, non-mechanical, physical process. This implied that a bottom-up description, framed in terms of the separate behaviours of constituents, would not do justice to the subtle and supple character of physical reality. The intrinsic unpredictability of chaotic systems is to be interpreted as leaving room for the operation of top-down organizing principles, which complete the description of what happens by their accounting for the way a system actually negotiates its labyrinthine envelope of possibility. The precise role of the additional openness represented by quantum mechanics' probabilistic operation was left unresolved in this picture. I have discussed elsewhere why this is necessarily so while we remain without an agreed understanding of how measurement takes place, for that understanding would be equivalent to a knowledge of how the microscopic and the macroscopic interlock.[22]

These higher-order principles act in a way corresponding to the input of information rather than energetic causation. That was the basis for the metaphysical hope, expressed in chapter 1, that this picture of physical reality might be the glimpse of the possibility of a dual-aspect monism, combining in a complementary way both mind and matter. But, of course, some sort of top-down causality must be operating, in appropriate ways, in very many systems, whose complexity falls far short of that required for the maintenance of consciousness or self-consciousness, and for the description of whose behaviour the category of the mental would not apply. I have suggested that this might be a locus, both of God's continuing interaction with his creation and also of his gift of freedom to his creation.[23] It is time to look a little closer at what these propositions might mean.

It is an important insight, clearly expressed by F. R. Tennant, that Christian theism 'must be sufficiently tinged with deism to recognize a relatively settled order'.[24] God's will is not whimsical. It is steadfast and he is the very antithesis of an arbitrary magician. Yet he is also personal and he can be expected to act in particular ways in particular circumstances. There

[20] Peacocke (1990), p. 121.
[21] Polkinghorne (1988), ch. 4.
[22] Polkinghorne (1991), ch. 7.
[23] Polkinghorne (1989a).
[24] Quoted ibid., p. 8.

is clearly some tension between these insights, but it can partly be resolved by supposing that the more intrinsically personal the circumstance, the more overtly personal God's response to it will be. Barbour, following some ideas of Tillich, has emphasized the need for the complementary use of both personal and impersonal models of God.[25]

To my mind, an important criterion of the personal is its uniqueness. Personal experience does not repeat itself – we hear a Beethoven quartet differently each time we listen to it, even if we play the same disc. On the other hand, impersonal experience is essentially repeatable, because it is relatively invulnerable to minor variations of context. Hence the possibility for experiment in science. Top-down information input which operates as a higher-order organizing principle in repeating circumstances will, in principle, be open to scientific study and description. I think it is entirely possible that the existence of the 'optimistic arrow' of increasing complexification in cosmic history (see p. 17), is understandable in terms of the operation of such (presently unrecognized) principles. Scientifically, they would move our understanding in the direction of a more holistic account of physical process than we presently possess. Bohm's implicate order is a conjecture in this spirit (pp. 23–4). Theologically, such organizing principles would be expressions of God's creative will exercised in the impersonal, 'relatively deistic' mode. Of course, theology already understands the known laws of nature in this sense. They are not the grain against which a wonder-working deity occasionally acts, but their regularities are the pale reflection of the faithfulness of the Creator.

The operation of such higher principles may well not exhaust the explanation of the remarkable fruitfulness of cosmic history. There may well have been throughout its unfolding a succession of particular critical points at which a divine influence was exercised in particular ways. If that is so, it would be scientifically indiscernible, veiled within the cloudy nature of the event in question. That would be Creatorly action in a more personal mode. I do not want to speculate much about how and when this happened. It would not be capricious, a kind of fitful poking by the divine finger, for God's care for his creation must be continuous, but it does not follow from that that there are not occasions when that care is exercised in specific ways. Divine faithfulness is not a kind of dreary uniformity. To illustrate the point, let me give one hypothetical example. Very early in the universe's history (if cosmological speculation be true) there was a sequence of events in the course of which the presently-experienced forces of nature crystallized out from the original, highly symmetric, grand unified state. The process is called spontaneous symmetry breaking and the balance between the strength

[25] Barbour (1974), pp. 84–91.

of the forces that results depends upon the infinitesimal (information-like) triggers which cause the crystallization to occur in this way rather than that. These force ratios are of anthropic significance. If they do not lie within certain narrow limits, the subsequent history of the universe will not be capable of producing carbon-based life. It is not inconceivable – I say no more than that – that part of the divine Creatorly activity brought it about that the ratios fell within the anthropic limits, at least in our observable universe.[26]

Many of those who write and think about science and religion will be uneasy at what I have just said. There is a curious ambiguity in their minds about the notion of *creatio continua*. They· eagerly embrace the general concept as providing a welcome theistic gloss on evolutionary history, but they are reluctant to acknowledge any actual divine activity within it. It seems to me that there is an implicit deism in much of what is said, whose nakedness is only thinly covered by a garment of personalized metaphor. Of course, I may well be mistaken in the detail of what I suggest – these matters are too far beyond currently established knowledge for anyone to claim certainty; pre-Socratic flailings around are all that any of us can manage – but I am unrepentant about the need to do the best one can. I have already given reasons for repudiating the charge that this is a return to the God of the gaps.[27] The point is that we are considering possibilities that arise from the intrinsically open character of physical process, not from transient patches of current human ignorance.

The presence of increasing complexity of structure as cosmic history unfolds, offers scope for the realization of more idiosyncratic and localized forms of top-down information input. That is ultimately our picture of the coming-to-be of consciousness and self-consciousness, the birth of the mental from the womb of the material. At an earlier stage of development, the presence of animal instincts would correspond to unconscious forms of this tendency.

God's gift of 'freedom' to his creation is conveyed by his respect for the integrity of these processes. In the case of inanimate creation, the outworking of these principles will not be overruled, though their effects may be reinforced in positively fruitful ways (as perhaps my spontaneous symmetry breaking example might illustrate). In the case of animate creatures, there is a much greater degree of autonomy to be respected, and I believe that God will interact with them in ways that are appropriate to their natures. That interaction culminates in the well-documented record of

---

[26] The symmetry breaking might have taken different forms in different cosmic domains. The anthropic condition requires, then, that one of them had the right form.
[27] Polkinghorne (1991), p. 46.

humanity's encounter with the divine in the depths of being. 'I have learnt to love you late, Beauty at once so ancient and so new! . . . You were within me, and I was in the world outside myself.'[28] Augustine met with God in the beautiful things of his creation, but the brightest illumination was to be found within.

Of course, there are delicate questions of balance between the need for divine respect for the integrity of creation and the need for divine scope for continuous interaction with it. As I have said already, this presents us with the classic theological perplexity of grace and free will, written cosmically large. One is trying to steer a path between the unrelaxing grip of a Cosmic Tyrant and the impotence or indifference of a Deistic Spectator. I have already indicated that I believe process theology to be impaled on the impotent branch of the horn of the dilemma.

Some interesting consequences flow from the adoption of the point of view that I am espousing. One relates to our understanding of the meaning of the phrase in the creed, 'heaven and earth'. The heaven referred to cannot properly be the place of our eschatological destiny, for that is the 'new heaven and the new earth' (Rev. 21.1) which we shall have to consider in a later chapter. I find Jürgen Moltmann's discussion of this question to be very helpful.[29] We are not stuck with an embarrassing remnant of an outdated cosmology (a realm above the blue dome of the sky), but instead we are offered a concept of symbolic richness. Heaven is the outward completion of earth, in the direction of the open and the unknown. A world without heaven would be a world without the possibility of transcendence, in which Berger's 'signals' would be illusory. 'A world like this would be a closed system, resting and revolving within itself. A world without transcendence is a world in which nothing new can ever happen. It is the world of the eternal return of the same thing.'[30] In other words, it would be the Laplacean world of a deterministic physics, in which there is rearrangement but no true novelty. The burden of my tale is that modern physics does not condemn us to such a world of frozen being. In the light of our discussion, when Moltmann says, 'We call the determined side of this system "earth", the undetermined side "heaven",'[31] we can give scientific encouragement to what he is driving at. One might venture the thought that earth is process read downwards towards the material, heaven is process read upwards towards the mental. As I have suggested before,[32] we mind/matter amphibians participate in a noetic world as well as in a physical world. The

[28] Augustine, *Confessions* (Penguin Books, 1961), 10.27.
[29] Moltmann (1985), ch. 7.
[30] ibid., p. 163.
[31] ibid.
[32] Polkinghorne (1988), pp. 75–80.

everlasting truths of mathematics are part of that noetic heaven. Moltmann suggests that God's potentialities and potencies are also to be found there, but that it is not the home of his being. Heaven is the arena of his economy (in the Patristic sense) but not the place of his essence, for heaven too is a created world. That is entirely consistent with our picture of the flexible openness of process being the locus of God's interaction with his creation, without there being an improper bridging of the ontological gap between the Creator and that creation. The concept of heaven and earth is intimately connected with the concept of *creatio continua*: 'for theology "heaven and earth" are not the petrified duality of a finished and completed universal condition. They are the two sides of the divine creative activity, the divine love and the divine glorification.'[33]

Another consequence of the picture I am proposing is that God interacts with the world but is not in total control of all its process. The act of creation involves divine acceptance of the risk of the existence of the other, and there is a consequent kenosis of God's omnipotence. This curtailment of divine power is, of course, through self-limitation on his part and not through any intrinsic resistance in the creature. It arises from the logic of love, which requires the freedom of the beloved. God's acquiescent will is part of every event, for if he did not hold the world in being there would be no such event at all, but his purposive will is not fulfilled in everything that happens. God remains omnipotent in the sense that he can do whatever he wills, but it is not in accordance with his will and nature to insist on total control. I also believe that by endowing his creation with the power of true becoming, God has permitted a kenosis of his omniscience, parallel to the kenosis of his omnipotence. Even he does not know the unformed future, and that is no imperfection in the divine nature, for that future is not yet there to be known. The adoption of this view requires us to take God's temporality very seriously.[34]

Such an account of God's relationship to his creation is highly contentious. On the one hand, there are many who feel it evacuates God of too much of his power. They maintain the contrary opinion that his hand is positively active in all that happens. Austin Farrer follows a neo-Thomist line in writing, 'God's agency must actually be such as to work omnipotently on, in and through creaturely agencies, without either forcing them or competing with them.'[35] Although Farrer wrote so extensively and so entrancingly about his notion of double agency,[36] I find it an unintelligible

[33] Moltmann (1985), p. 164.
[34] Polkinghorne (1989a), ch. 7.
[35] Farrer (1968), p. 76; see also White (1985).
[36] ibid., and Farrer (1967).

kind of theological doublespeak. On the other hand, much the more common sort of modern reaction has been to recoil from any notion of particular divine agency and restrict God's role to that of simply holding the world in being. Such a view is taken by Maurice Wiles. He summarizes his position by saying that

> we can make best sense of this whole complex of experience and of ideas if we think of the whole continuing creation of the world as God's one act, an act in which he allows radical freedom to his human creation. The nature of such a creation, I have suggested, is incompatible with the assertion of further particular divinely initiated acts within the developing history of the world.[37]

Supporters of this point of view sometimes deny that they are deists, on the grounds that the single great act of which they speak is a timeless upholding and not a mere initiation of cosmic history, but the God of their account is certainly not a personal God, able to react in particular ways to particular occurrences.

There seem to be two motives for being willing to settle for so minimal a view of divine agency. One is a feeling that modern science will permit us nothing more. I hope I have already shown that to be a mistaken opinion. The other, and much the more significant, reason is the desire to solve the problem of theodicy by absolving God from any possibility of responsibility for specific happenings. A God of the 'one great act' is not a God who can be blamed for the Holocaust. Yet he is also not the God who raised Jesus from the dead. I have already emphasized the remarkable phenomenon of hope (pp. 65–7); our account of divine agency will have to be adequate both to the fact of evil and to the fact of hope.

The problem of theodicy is easily stated. How can a world of cancer and concentration camps be the creation of a God at once all-powerful and all-good? Jung sought to solve the problem by modifying the idea of God's goodness. He speaks of a dark side to the divine nature.[38] But such an ambiguous God could not be the reliable ground of hope. Christian thought has preferred to reconsider what is meant by 'all-powerful'. The kenosis of the creative act is then held to afford the necessary insight.[39] Keith Ward characterizes the dilemma we face: 'It often seems that we can neither stand the thought of God acting often (since that would infringe our freedom), nor the thought of him acting rarely (since that makes him responsible for our

---

[37] Wiles (1986), p. 93.
[38] Jung (1954).
[39] See Polkinghorne (1989a), ch. 5.

suffering).'[40] We do not want to consider ourselves as part of God's puppet theatre, but if he acts to heal some people, why does he not act to heal many more? Pious tales of providence often seem so ludicrously out of scale with the immensity of the sufferings of the world. 'Moral considerations lead us to prefer to deny that God can act at all, rather than to say that he acts arbitrarily or manipulatively.'[41]

There is great mystery in suffering, and I do not want to suggest that there is any facile way to understand its incidence. Yet I do not think we have to choose between a God who is inactive or arbitrary or (worst of all) a cruel manipulator. I have written before, 'there is only one broad strategy possible for any theodicy. It is to suggest that the world's suffering is not gratuitous but a necessary contribution to some greater good which could only be realized in this mysterious way.'[42] In relation to moral evil (the chosen cruelties of humanity), this leads me to embrace the free-will defence: that despite the many disastrous choices (and one cannot say that in this century without a quiver in one's voice), a world of freely choosing beings is better than a world of perfectly programmed automata. I have gone on to suggest that in relation to physical evil (disease and disaster) there is a parallel 'free-process defence': that in his great act of creation, God allows the whole universe to be itself. Each created entity is allowed to behave in accordance with its nature, including the due regularities which may be part of that nature. 'God no more expressly wills the growth of a cancer than he expressly wills the act of a murderer, but he allows both to happen. He is not the puppet master of either men or matter.'[43]

Two contrasting criticisms can be made of this proposal. One is to deny the reality of physical *evil* at all. Rather, the raggedness and malfunction that we label in this way are simply the necessary way things work out in an evolving universe, free to make itself by the shuffling explorations of happenstance, and they are fully intelligible and acceptable on those terms. Here, what was suggested as a defence in the face of a problem is being claimed, in fact, to be so successful that the problem is abolished outright. Process thinkers often take this view. Pailin says that understanding what is involved in creativity

> exposes the supposed 'problem' of natural evil to be an illusion . . . [some of the things that happen] may be unpleasant and need to be avoided or eradicated as far as possible in the interests of human flourishing but that they are here at all with their particular characteristics is no more God's

[40] Ward (1990), p. 2.
[41] ibid., p. 3.
[42] Polkinghorne (1989a), p. 63.
[43] ibid., pp. 66–7.

fault than it is due to divine intention that we exist with the capacity to see and hear within certain wavelengths.[44]

Could one put that to someone who had lost their family in an earthquake, or whose child was dying of leukaemia? Obviously, no one would suggest that one could. But it *would* be suggested that the tragedy in those circumstances lies in the human perception of them, our deep sadness at diminishment and transience. I would want to follow Langdon Gilkey in claiming that our human creaturely status as beings made in God's image (Gen. 1.26) implies that this human intuition is true and is to be taken with utmost seriousness,[45] and so these events are correctly to be seen as evil (and their precursors likewise, before self-consciousness arose in the universe), and they therefore do present a problem for theodicy. Peacocke says that 'from a purely naturalistic viewpoint, the emergence of pain and its compounding as suffering as consciousness increases seem to be inevitable aspects of any conceivable developmental process that would be characterized by a continuous increase in ability to process information coming from the environment'.[46] In a striking phrase, Rolston calls this 'cruciform naturalism'.[47] However intelligible this cost of complexity may be, that recognition alone does not seem to me to dispose of the problem. It needs addressing at a level more profound than the purely naturalistic.

The second criticism takes a somewhat different line. It questions the validity of the defence, alleging that it improperly imports notions of 'freedom' into an impersonal realm where they have no validity. There may be value in God's allowing a man or woman to make their autonomous decision, but what is the value in his allowing a tectonic plate to slip or a cell to become cancerous? Could not these be corrected without doing violence to the integrity of creation?

I take it we are not considering the possibility of a merely magical world in which fire changes its characteristics when it comes within a few inches of a human hand. There are familiar arguments that a rational God would not create so capricious a universe, and that morally responsible beings could not properly act within its shifting circumstance, in which all true consequence would be abolished. I take these arguments to be convincing. We are considering, therefore, a rational and orderly world. We all tend to think that had we been in charge of its creation, we would somehow have contrived it better, retaining the good and eliminating the bad. The more we

[44] Pailin (1989), p. 153; but see Pailin (1992) for a more nuanced account arising from experience.
[45] Gilkey (1959), p. 186.
[46] Peacocke (1990), p. 68.
[47] Rolston (1987), p. 289.

understand the delicate web of cosmic process, in all its subtly interlocking character, the less likely it seems to me that that is in fact the case. The physical universe, with its physical evil, is not just the backdrop against which the human drama, with its moral evil, is being played out, so that the two can be disentangled. We are characters who have emerged from the scenery; its nature is the ground of the possibility of our nature. Perhaps only a world endowed with both its own spontaneity and its own reliability could have given rise to beings able to exercise choice. I think it is likely that only a universe in which we could entertain a free-process defence, would be one in which there could be people to whom the free-will defence could be applied.

That is about as far as we can carry our consideration of the problem of evil in general terms. It is the most perplexing of all the difficulties that confront the religious believer. The distinctive Christian address to the issues it raises centres on the cross of Christ, seen as a divine participation in the brokenness and pain of the created order. The cross also radicalizes all notions of what could be meant by providence. Timothy Gorringe writes:

> What makes God the world Ruler, said Barth, as opposed to all false gods and idols, is 'the very fact that his rule is determined and limited: self-determined and self-limited, but determined and limited none the less'. Knowledge of this self-limitation derives from the cross, for if everything were rigorously determined what could the cross be but a piece of spectacular, though indecent, theatre? On the contrary, the 'necessity' of the cross, frequently spoken about by New Testament authors, is God's refusal to overrule human history. If the cross is our guide, God is no determinist.[48]

Gorringe says that 'Providence is God's wisdom in action, and what that means is seen in Christ,'[49] and he goes on to expound this in terms of the cruciform wisdom of 1 Corinthians 1.21–5. His conclusion is that 'the cross is the heart of providence'.[50]

Yet the Christian cannot think of the cross without also thinking of the resurrection. A crude 'pie in the sky' theodicy, which simply justifies the sorrows of earth by appeal to the prospective joys of heaven, will not do, but neither will it do to neglect the implications of the Christian hope of a destiny beyond death. I shall take up questions of ultimate fulfilment in chapter 9.

The doctrine of creation receives comparatively little attention in

[48] Gorringe (1991), p. 12.
[49] ibid., p. 55.
[50] ibid., p. 103.

CREATION

contemporary theology. Hans Küng could write a six-hundred page account of Christianity for the general educated reader without creation occurring even once in the index.[51] Theological discourse has become unduly human-centred, and Pannenberg is right to say that Christian theology 'by limiting its frame of reference to man has allowed Christian belief in creation to atrophy'.[52] In fact, humanity and the physical world that gave us birth belong inextricably together. The appearance of self-conscious beings has profoundly modified the course of evolutionary history, for now there is an alternative mechanism, of great power and effectiveness, for transmitting information from one generation to the next, other than by coding in DNA. Natural selection has been modified. We do not leave the weak to die. The effect of culture in the few thousands of years in which it has been operating has already produced very remarkable results, even to the point at which the accumulated skills of human knowledge provide the possibility for culture to intervene in the genetic process itself, by means of the ambiguous gift of genetic engineering. This phenomenon of human influence upon the course of the natural world has led Philip Hefner to speak of humanity as the 'created co-creator',[53] an idea which Peacocke reminds us goes back at least as far as the confident days of the Christian humanists of the Italian Renaissance.[54] Whatever one makes of that particular form of expression – and I prefer more modest, stewardly language – the thought recalls to us our human responsibility for the integrity and sustainable fruitfulness of the world which we inhabit. These 'green' issues are of great importance for the future of the Earth, but there is not an opportunity here to do more than glance at them. A properly understood doctrine of creation is an important undergirding of our intuition of the value of the world, quite apart from that world's utility for our own purposes.[55] Speaking of the mythical monster who is a symbol of the non-human order, the Lord says to Job, 'Behold, Behemoth, which I made as I made you' (Job 40.15).

I do not think, however, that this respect for creation need find expression in a Jain-like refraining from any interference with other forms of life. There are natural enemies of humankind who are to be resisted. A good deal of shrewd sense is contained in the couplet:

> He loveth all, who loveth best;
> The streptococcus is the test.

[51] Küng (1977); see also the survey of Macquarrie (1981) where there are sixteen references to existentialism in the index but only four to creation.
[52] Pannenberg (1976), pp. 26–7.
[53] In Peters (1989), ch. 6.
[54] Peacocke (1979), p. 305.
[55] cf. Bradley (1990).

We must be careful not to allow sentimentality to distort our perception of the reality of the living world. There are uses to which animals may legitimately be put, including their consumption as food, if they are humanely carried out. Where I think the Christian should protest is where an animal's nature is frustrated and diminished by human intervention to a grievous degree. The intolerably close-confinement of some factory-farming methods seems hard to justify. I personally do not feel a similar difficulty about many forms of hunting, which seems to me to be a much more natural activity. Perhaps I am influenced by my maternal grandfather, one of the kindest of men. He was a head groom and frequently rode to hounds as a second horseman. He cared for animals deeply but felt no incongruity in his life.

In many perplexing ethical decisions there is a balance to be struck between competing possibilities for good. I accept the need to use animals in carefully controlled medical research, but we must be more critical of their use in routine testing procedures for cosmetics.

The dangers arising from a scale of human activity capable of affecting the terrestrial environment are to be taken seriously, though they are indeed hard to evaluate. The subtle complexities of ecological feedback make the predictions of models very uncertain in their relevance, however confidently they may be proclaimed. This applies, of course, to those forecasts which are optimistic as well as to those which are pessimistic. What does seem certain is that the politically very delicate question of population control is central to the attainment of a sustainable strategy.

I feel a certain ambiguity about current concerns with green problems. I have acknowledged the importance of the questions, but we must be careful that the discussion is balanced and not shrill, the approach realistic and not distorted. The Christian will not want to cry peace where there is no peace, nor war where there is no war. A prophetic vision will be a clear vision, clouded neither by utopian longing nor by sentimental fancy. As part of that clarity we shall want to perceive the strangeness of nature, its own prodigalities and ruthlessnesses. We shall need to behold Behemoth, even as we remember that the Lord made him as he also made us.

5

~~~~~

Jesus

'One Lord, Jesus Christ'

More than half the creed is devoted to the paragraph concerning Jesus Christ. It is a commonplace to say that Christianity is the most historically orientated of the world's religions, and the focus of that historical concern is the life of Jesus of Nazareth. Almost all that we may know of that life must be derived from the writings of the New Testament; all that could be gleaned elsewhere is the mere indication that such a person existed and that he was executed. The development of the critical study of history since the Enlightenment, with the consequent quizzical, even sceptical, approach to historical sources, has presented to Christian thinking a fundamental challenge of reassessment. No longer is it possible for us simply to read the gospels as plain, matter-of-fact accounts of the words and deeds of Jesus, such as might have been recorded by an unusually careful reporter of the day. They are something more subtle than that, and their study requires a corresponding sophistication.

Immediately a problem presents itself. The books of the New Testament in general, and the gospels in particular, must be the most researched and continually reassessed writings ever submitted to scholarly scrutiny. For more than two hundred years, a great industry of learning has busied itself with that critical analysis. I like to read as much as I can of that kind of writing but, of course, I can claim no expert status in speaking on these issues. How then can I have the temerity to address the questions? One answer would be that I do not have any choice. A Gifford Lecturer must be a bottom-up thinker. He must start with the phenomena – and the foundational phenomena of Christianity are set out in the New Testament. But it is not mere necessity which drives me to the task. While I respect and value the insights that scholars provide, I cannot believe the matter should be left solely in their hands. If we were presented with a substantial body of 'agreed results', that submission to scholarly authority might be possible. But the fact of the matter is that we are not. A survey of thought about the New Testament[1] soon reveals the clash of view and the ebb and flow of fashion

[1] See, for example, Neil (1964).

so characteristic of any scholarly activity in which, through successive generations working with limited primary sources, each of those generations seeks to establish its own originality. I do not for a moment deny the truthful intent of this labour, or that each generation succeeds in some way in deepening our understanding, but I am conscious of the effect of the pressing academic need to say something new (so familiar in my own field, theoretical physics) and also of the special problems in this area resulting from our inability to agree on widely accepted criteria by which results could be validated.

This lack of agreement derives, I believe, partly from the significance of the issues at stake. Historical judgements inevitably call for a high degree of the use of those tacit skills whose exercise we have already recognized as being an essential part of the rational pursuit of knowledge (chapter 2). It is not surprising, therefore, that there should be differences of opinion about which gospel sayings are authentically to be attributed to Jesus. What is surprising is the degree of such disagreement. Ancient historians have sometimes criticized New Testament scholars for their marked reluctance to trust their sources.[2] But in many people's minds, much more hangs on the assessment of the life and character of Jesus than hangs on the assessment of the life and character of Julius Caesar. What we think about him *matters*. One result of this is to demand in the former case a greater degree of scrupulosity than is required in the latter case; another result is that in the case of Jesus a hidden agenda of wider significance operates to influence the mode of ostensibly historical judgement. This hidden agenda can generate either an undue scepticism or an undue literal reliance.

In the end I do not think one can divorce New Testament scholarship from Christian theology, any more than one can divorce experiment from theory in science. The current academic habit of respectfully observing such a division in the theological world has not been helpful. We need more people bold enough to venture across scholarly boundaries. I think that extends even to amateurs like myself, who have a certain right to talk about the general character of the wood, even if they do not have the expertise correctly to identify every one of its individual trees. Without further apology, I direct myself to the task.

The gospels are an idiosyncratic sort of writing. Their concern is with Jesus of Nazareth, but they are clearly not biographies in a modern manner. Not only do they fail to record all sorts of things that we, with our modern concerns, would rather like to know (what did he look like?), but they are clearly written from a point of view and for a particular purpose. The very

[2] Sherwin-White (1963).

word *euangelion*, asserted in the first verse of the earliest gospel[3] (Mark 1.1), means 'good news' and, for his part, the author of the fourth gospel is clear what he is about: 'these are written that you may believe that Jesus is the Christ, the Son of God, and that believing you may have life in his name' (John 20.31). Yet the form of this good news is the story of a life and death and its aftermath. There is a clear intent to root the gospel in the events of history.[4]

We are familiar with the distinction in character between the three synoptic gospels and the fourth gospel: the synoptic Jesus, uttering pithy sayings and telling pungent parables, a first-century Palestinian proclaiming the kingdom of God; the Johannine Jesus, speaking in timeless tones, whose long discourses concern, not the kingdom, but himself, proclaimed as the Good Shepherd and the True Vine. Doubtless there is a mingling in John of the historic Jesus and the post-resurrection exalted Christ, and perhaps that admixture is not wholly absent from the other gospels either, but we must remember that John chose to write a *gospel* (that is, an account centred on a life and not just a series of timeless meditations), and that in geography and chronology we have good reasons for taking his version very seriously.[5] History is closely woven into the fourth gospel (e.g., archaeological research has shown that the five porticoes of the pool by the sheep gate (John 5.2) were actually there and they are not a symbol for the five books of the *Torah*, as some commentators had fancifully supposed).

There is neither a pedestrian literalism nor a cavalier disregard in the gospels' attitude to historical events. It seems to me that the writers are neither slavish about detailed accuracy nor careless about what actually happened. As to the first point, it is instructive to compare two versions which most of us would think originate from the hand of the same author. Luke, in his gospel, claims to write an 'orderly account' (Luke 1.3), and in Acts, by implication, to continue it (Acts 1.1), but it does not seem to trouble him that in the first book Jesus parts from his disciples on the first Easter day (Luke 24.51; taking the longer reading of the text) while in the second book this parting is dated forty days later (Acts 1.3). Of course, the ascension is a peculiar kind of event in which the symbolic predominates over the literal (see p. 123), but one sees a rather similar looseness of chronological attribution in the well-known conflicts between the synoptics and John as to the dating of the cleansing of the Temple and the relation of the crucifixion to the day of Passover.[6]

[3] I follow conventional thinking about questions of dating New Testament documents; but see Robinson (1976) for different views.
[4] cf. Wright (1992), especially pp. 396–403.
[5] Dodd (1963); Robinson (1985).
[6] Many scholars think that John knew Mark's gospel.

On the other hand, there are many indications in the gospels of a respect which the authors had for elements of the historic tradition, even when they were puzzling or embarrassing from the perspective of those writing a generation or two after the events themselves. Mark records Jesus as replying to a man who addressed him as 'Good Teacher', by saying, 'Why do you call me good? No one is good but God alone' (Mark 10.18). Not a remark you would expect to be perpetuated about a revered leader, unless he had actually said it. Matthew felt the difficulty and toned it down into the innocuous, 'Why do you ask me about what is good?' (Matt. 19.16), though Luke was made of sterner stuff and retained the saying (Luke 18.19). However, Luke apparently could not stomach that desolate word from the cross: 'Eloi, Eloi, lama sabachthani? . . . My God, my God, why hast thou forsaken me?' (Mark 15.34), and Matthew turned the Aramaic into Hebrew (Matt. 27.46), so that it became a scriptural quotation rather than a cry of dereliction.

Then there are shocking things that Jesus said: 'Leave the dead to bury their own dead' (Matt. 8.22, par.), a remark in flagrant violation of Jewish sacred duty to parents and, indeed, of all pious opinion in the ancient world,[7] and thus gratuitously offensive to readers of the gospel. There are the difficult verses in which Jesus is recorded as implying a fulfilment in the lifetime of his hearers which, in the most immediate sense of his words, did not come about in any obvious fashion (Mark 9.1; 13.30, par.). When Matthew wrote of Jesus as sending out the twelve with the words, 'You will not have gone through all the towns of Israel, before the Son of man comes' (Matt. 10.23), the evangelist was as aware as we are that that saying did not find an obvious literal fulfilment. Yet he put it in his gospel.

To these examples must be added some baffling verses in the gospels which seem to be there for no reason other than that they must have formed part of an actual reminiscence whose significance is now partly lost to us. Cases would be the young man stripped naked in flight (Mark 14.51–2) and that odd conversation in which Nathanael is so impressed by Jesus' saying that he saw him under a fig tree (John 1.48), an impression which arises for reasons which now altogether escape us. The commentators' remarks that in this incident Jesus is portrayed as exhibiting a miraculous power of sight-at-a-distance seem to me to be peculiarly lame.

These internal considerations of the gospel texts encourage in me the view that their authors were people who were attempting to achieve a genuine historical authenticity in their accounts of Jesus. They were concerned to tell it like it was, within the conventions of their time. They would have been making use of the oral tradition transmitted in the

[7] Sanders (1985), pp. 252–4.

generation that intervened between the crucifixion and the writing of Mark (a gap equivalent to that between the present day and my own first encounters with leaders of the high energy physics community, of which I believe I retain a lively and essentially dependable reminiscence). The world of their day was one which cultivated the art of oral transmission, and there are reasons to suppose that there might also have been written records now lost, such as the hypothetical, but plausible, Q, still held by many scholars to lie behind the sayings material common to Matthew and Luke.

Of course, one does not deny that the evangelists moulded their material, by selecting it and ordering it. It is instructive to compare the versions of the Lord's Prayer (Matt. 6.9–13; Luke 11.2–4) and the Beatitudes (Matt. 5.3–11; Luke 6.20–3) presented by two gospel writers. There is basic agreement on substance but significant variation of emphasis and elaboration. Redaction criticism has cast helpful light on this process.[8] Nor can one deny that the early Christian communities sometimes felt able to create prophetic words which spoke to their circumstances and which were then incorporated into the tradition as if they had been uttered by the earthly Jesus. A passage like Matthew 18.15–17, which gives detailed instructions about how to deal with sinners in the 'church' (*ecclēsia* only occurs here and in Matt. 16.18 in all the gospels) quite obviously arose in this way. Yet there were equally obviously limits to this process. Acts and the Pauline epistles make it clear how critical a question it was in the early Church whether Gentiles should be circumcised or not, but no one invented a word of Jesus to settle the issue. Paul's discussion of marriage (1 Cor. 7.8–16) makes it plain that he distinguishes between what he believes originates from the Lord and what is his own apostolic judgement. Sometimes one may suppose that development took place which drew out the implications of something Jesus had said, amplified in the light of further experience. He can scarcely have been as explicit about food regulations as Mark 7.14–23 par. suggests – for otherwise how could the disputes have arisen, to which Acts and the Pauline epistles testify? – and clearly Mark 7.19b ('Thus he declared all foods clean') is an editorial gloss underlining the message. Yet it is not inconceivable that the original enigmatic saying about defilement coming from inside rather than outside was in fact Jesus' utterance.

Scholars have sometimes sought to sift the words of Jesus from later constructions by having recourse to the criterion of 'double dissimilarity'. A saying is considered authentic if it expresses a view distinct both from contemporary Jewish thought (as far as that is known) and from the concerns of the early Church. While one can concede the positive force of this criterion, to use it as the sole test of authenticity is clearly absurd. It

[8] See, for example, Sanders and Davies (1989), ch. 14.

would produce a Jesus entirely without anchorage in his society and bereft of lasting influence upon his successors. Applied to a scientist, it would assign to Schrödinger his idiosyncratic and unsuccessful attempts at unified field theories but regard his celebrated quantum mechanical wave equation as being of questionable attribution.

It seems to me that questions of authenticity cannot be settled simply by the application of quasi-algorithmic evaluative procedures, like double dissimilarity, but that one has to take the risk of relying on the tacit skill of judgement. Biblical studies must resist the temptation to devise procedures which ape a superficial notion of scientific method. Rather, it should have the intellectual nerve to proceed in its own proper way. This will require a combination of critical assessment with empathetic interpretation. I cannot feel altogether happy with Sanders and Davies' assertion that 'The basic means of establishing evidence is cross-examination. The gospels must be treated as "hostile witnesses" in the court room.'[9] It seems to me that we must take the risk of a more subtle and open approach.

The first question, surely, is whether the gospels give the impression of having behind them a powerful personality whose character we can, at least partially, discern. I believe the answer to be yes. Just to take one example, consider the question of the parables. It is extraordinarily difficult to make up tales which have the penetrating power and haunting quality which the stories of the prodigal son (Luke 15.11–32), or the good Samaritan (Luke 10.29–37), or the parable of the sheep and goats (Matt. 25.31–46) possess. Yet some scholars are willing to suppose those small early Christian communities to have been well-endowed with creative story-tellers of this remarkable kind. It seems to me much more likely that there is one outstanding mind behind most of the parables. Once again, that is not to deny the subsidiary role of the early Church in moulding and using the material. The reference to the nations (*ethnē*) in Matthew 25.32 may well originate in the context of a Gentile mission, rather than the ministry of Jesus, which seems well-attested as being directed towards Israel.

The acuteness in controversy, by which the questioner so often has the question turned round to reveal his own presuppositions and consequent prejudice (see, for instance, healing on the sabbath (Mark 3.1–6, par.), authority (Mark 11.27–33, par.), and the tribute money (Mark 12.13–17, par.)), again seems to me to be the record of the action of a distinct and incisive mind.

Thus I find the agnosticism of some scholars about what can be known of Jesus to be most unconvincing. Talk of the 'largely unknown man of

[9] ibid., p. 301.

Nazareth'[10] seems to me to be grotesque. I would rather say with Pannenberg that 'It is quite possible to distinguish the figure of Jesus himself, as well as the outlines of his message, from the particular perspective in which it is transmitted through this or that New Testament witness.'[11] The considered judgement of C. H. Dodd was that 'the first three gospels offer a body of sayings on the whole so consistent, so coherent, and withal so distinctive in manner, style and content, that no critic should doubt, whatever reservations he may have about individual sayings, that we find reflected here the thought of a single, unique teacher'.[12] Let me add one more voice to show that I take my amateur stand in the company of scholars of great learning and unquestioned integrity. Charlie Moule summarizes a study of the tradition by saying:

> the general effect of these several more or less impressionistic portraits is to convey a total conception of a personality striking, original, baffling yet illuminating. And it may be argued that it is difficult to account for this except by postulating an actual person of such a character. The very fact that the total impression is made up of several different strands of tradition, originating (one may reasonably presume) in different circles, compensates in some degree for the absence of any rigorous test by which the authentic and original has been isolated within any one strand of tradition. If all of them, for all their diversity, combine to create a coherent and challenging impression, this is significant.[13]

Yet I must also concede that there are aspects of the gospel picture which are uncongenial to the modern reader. We need not accept Albert Schweitzer's exaggerated depiction of an apocalyptic figure trying to force a divine turning of the wheel of history by his self-immolation, but the note of urgent choice in the face of the imminent arrival of a new age is certainly there (Mark 10.29–30, par.; Matt. 8.11–12, par.; etc.), as is the idea of an eschatological trial prior to the End (Mark 10.38, par.; Mark 13, par.). Though these themes may have been elaborated by the early Christians, it is difficult to think that something like them was not also part of Jesus' view. He is thoroughly embedded in the thought patterns of his age. If he were truly human, what else could he be? Yet his concern is with God's judgement and mercy, not with datable prediction (Mark 13.32, par.).

Much more disturbing are those passages in which Jesus is portrayed as speaking in scathing, or even vituperative, terms of his opponents (Mark 12.38–40, par.; Matt. 23.13–36; etc.). Perhaps some of the language reflects

[10] Hick (1976), p. 117.
[11] Pannenberg (1968), p. 23.
[12] Dodd (1971), p. 33.
[13] Moule (1977), p. 156.

the later tension between church and synagogue, but we must be wary of making hard sayings easy by attributing all the disagreeableness to the early Christians. There must have been a forceful sternness about Jesus which makes the cleansing of the Temple a credible episode. There is a puzzling complexity, as well as an attractiveness, about the figure of Jesus which drives those who speak of him to use words like 'baffling' or 'mysterious'.

One final general point must suffice. The New Testament writings are, on anyone's view, a remarkable body of religious literature. The Christian Church which has sprung from them has proved a persistent and fruitful movement, however ambiguous aspects of its history may be. These are notable phenomena, whose earliest witnesses consistently relate their experience to an origin expressed in terms of Jesus. To say the least, it would seem reasonable to explore the possibility that in this man is to be found the clue to the interpretation of what followed him, rather than locating it in the supposed creativity of a community which somehow conjured such richness from the obscure recollection of a dim and shadowy figure preceding them. In Martin Hengel's words, 'Even the most radical sceptic cannot avoid the simple historical question how this simple wandering teacher and his outwardly inglorious death exercised such a tremendous and unique influence that it still remains unsurpassed.'[14]

Albert Schweitzer undermined the efforts of the nineteenth-century authors of 'biographies' of Jesus by showing how each had constructed a figure in his own image. There followed the period of the ascendancy of Rudolf Bultmann and the form critics, in which the early Church received all the credit for insight and originality. Nerve was recovered with the 'new quest' of the historical Jesus, initiated by Ernst Käsemann. Let us see what we can make of this task.

One of the points on which there is most scholarly agreement is that Jesus proclaimed the kingdom or rule of God: 'The time is fulfilled, and the kingdom of God is at hand; repent, and believe in the gospel' (Mark 1.15). His words and deeds are part of that manifestation (Matt. 12.28, par.; Matt. 13.16–17, par.; etc.). Many of the parables, including some otherwise hard to understand (e.g., the unjust steward, Luke 16.1–8), seem to have as their point the urgency of responding to this outbreaking of divine rule in the end-time of history. The offer of the kingdom is there for those able to receive it: 'Blessed are you poor, for yours is the kingdom of God' (Luke 6.20, par.). The New Testament usage of the phrase is overwhelmingly contained in the synoptic gospels, which strongly suggests its authenticity on the lips of Jesus.

E. P. Sanders says, 'It used to be almost a taboo question in many circles to

[14] Hengel (1981), p. 72.

ask what Jesus was up to.'[15] He believes this question is answerable and that one should start with some of the acts of Jesus, which Sanders thinks can be established with greater certainty than his words. 'Almost indisputable facts' are that Jesus was baptized by John, that he was a Galilean who preached and healed, called disciples and spoke of a group of twelve, confined his activity to Israel, was involved in controversy about the Temple and was crucified outside Jerusalem by the Romans. To these facts Sanders adds that the movement around Jesus persisted after his death and that it was to some extent persecuted by the Jews.[16] At the end of his discussion Sanders gives a list of further conclusions ranging from the 'certain or virtually certain' to the 'conceivable', which place Jesus within the Jewish contemporary setting as Sanders describes it (broadly 'restoration eschatology', centring on the hope of a new Temple, and 'covenantal nomism', the community of grace within the Law), while emphasizing that Jesus took an unprecedented line in accepting sinners within the kingdom of God without prior insistence on repentance. Sanders also adds a list of some 'incredible' conclusions, rejecting any idea which would make the Pharisees implacably legalistic and give to Jesus a monopoly of the concepts of love and mercy.[17] It is clear that a perspective is here imposed on the selection of gospel material to accord with Sanders' learned reconstruction of first-century Palestinian Judaism.[18]

John Macquarrie has produced a list of seven or eight items of information about Jesus which can be culled from Paul's writings. They range from the comparatively trivial (Jesus had brothers, 1 Cor. 9.5) to the highly significant, like the institution of the Eucharist (1 Cor. 11.23–6), and – the bracketed eighth 'fact' – that he was raised on the third day (1 Cor. 15.4).[19] Of course, one might expect such a collection, stemming from occasional writings like the Pauline letters, to have an assorted and rather odd character to it. Later Macquarrie compares this list with Sanders' and with a list of characteristics of similar length conceded by Bultmann (and, in contrast with Sanders, placing considerable emphasis on a confrontation with Jewish legalism[20]). He asks, 'Does a meagre list of the kind we are considering tell us enough about the "what" of Jesus to make the "that" identifiable, to give it some substance so that it is not just a vague assertion that someone existed and was crucified, but we cannot say anything specific about that "someone"?'[21] My answer would be that it is not enough.

[15] Sanders (1985), p. 2.
[16] ibid., p. 11.
[17] ibid., pp. 326–7.
[18] See also Sanders (1977).
[19] Macquarrie (1990), pp. 51–2.
[20] cf. the comparison of Jesus and the Pharisees in Harvey (1982), pp. 50–1.
[21] Macquarrie (1990), pp. 350–1.

Macquarrie likes to talk about 'the Christ event', that is to say the pre-Easter and post-Easter phenomena which together constitute the seminal experience which gave rise to Christianity. One apparent attraction of this approach is that it enables one to decline too close an investigation into what originates in the early Church and what is to be attributed to the historical Jesus. But without the pre-Easter there could have been no post-Easter – and that includes the resurrection itself (to be considered in detail in the next chapter), which could not have been merely a whimsical divine decision to exalt an obscure Galilean preacher, unjustly executed, but rather its fittingness must have been deeply rooted in exactly who Jesus was. I cannot myself rest content with an account of the historical pre-Easter Jesus which falls short of an intelligible basis for all that is claimed about Easter and its aftermath.

We return to the gospels, therefore, seeking from them more than a skeletal, minimal portrait, and with the expectation that it should prove possible to detect, in Moule's words, 'a personality striking, original, baffling yet illuminating', who might indeed credibly have been the trigger for the great Christian explosion into history.

I think that a good way into the quest is through the attempt to identify traces of the *ipsissima vox* of that striking personality, looking for tones and phrases which by their vividness suggest the impress of a well-remembered, oft-recounted character.[22] I have already suggested that the haunting nature of some of the parables, and the penetrating challenges to prejudged suppositions, carry for me just that note of encounter with a highly original and individual mind. Another clue is provided by those untranslated Aramaic words which occasionally turn up in the Greek text of the New Testament. Much the most frequent of these is *amen*. First-century Jews would have used this as we do today, at the end of a prayer to signify assent. Jesus' usage, however, is entirely different. He places it at the beginning ('Amen, I say to you . . .'), where its force is to assert the unshakeable certainty of what is to follow. This occurs frequently in all four gospels – in John, in the even more emphatic reduplicated form 'Amen, amen . . .' It is a pity that so many versions disguise this fact by using translations such as 'truly', which fail to acknowledge the singularity of the underlying word. I personally think that this idiosyncratic usage is a preserved reminiscence of a characteristic form of speech and that it testifies to one who was conscious of possessing a particular and true insight, so that he 'taught them as one who had authority, and not as the scribes' (Mark 1.22, par.).[23] It is indeed believable that he spoke as Matthew portrays him in the Sermon on the

---

[22] Jeremias (1971), ch. 1.
[23] ibid., pp. 35–6.

Mount (Matt. 5), saying, 'You have heard that it was said to the men of old
. . . But I say to you . . .' It is the *Torah*, the Mosaic Law, which is being
deepened or even corrected (Matt. 5.32, 39, 44) in these passages. Here is no
rabbinic commentator, offering an interpretative gloss, but instead someone
who is claiming for himself an authority capable of being set alongside the
divine revelation from Sinai. The formula is not even the prophetic, 'Thus
says the Lord . . .', but, 'This is what I say . . .' The implications of that bear
some pondering. Jesus is also portrayed as exercising an authority, perceived
by critics as amounting to the usurping of divine prerogative, when he
pronounced the forgiveness of sins (Mark 2.5–9, par.; Luke 7.48–9).

Another significant Aramaic word is *abba*, an intimate family word for
father, with more than a trace of 'daddy' to it, used as an address to God.[24]
It only occurs once in the gospels, in the fraught setting of Jesus' prayer in
the Garden of Gethsemane (Mark 14.36), where it is immediately translated
'Father' (*patēr*). The retention of the Aramaic together with the Greek
suggests that here is another significant reminiscence, an impression
strengthened by the continued use of *abba* in the early tradition (Rom. 8.15;
Gal. 4.6). It would seem to speak of a peculiarly close relationship that Jesus
experienced with God, something altogether deeper and more direct than
that expressed in the respectful and more distant address which would have
been contemporary Jewish practice.

The next phrase to consider is one on which there has been perhaps more
discussion and disagreement than almost any other in the New Testament. It
is 'the Son of man'. Although the equivalent Aramaic phrase (*bar nash(a)*)
never appears transliterated in the New Testament, the expression is so odd
in Greek (as odd, in fact, as it is in English) that it is clear that it is an
Aramaicism which is being reproduced. Three kinds of consideration have to
be taken into account in seeking an understanding of the New Testament
usage.

Firstly, 'son of man' is a perfectly normal semitic way of articulating what
we would more simply express through the single word 'man'. The Old
Testament furnishes many examples of this use of the equivalent Hebrew
phrase *ben adam* (e.g., Ps. 8.4 and Ezekiel *passim*). In the New Testament,
however, the predominant usage is to employ the definite article in the
Greek, encouraging the translation 'the Son of man'. This fact has been
emphasized particularly by Moule,[25] who takes it as indicating that
reference is being made to a particular figure, rather than humanity in
general. The natural referent is the 'one like a son of man' of Daniel 7.11–18,
a heavenly figure involved in the vindication before God of the persecuted

[24] ibid., pp. 61–8, but cf. Vermes (1983), pp. 210–13.
[25] Moule (1977), pp. 11–22.

'saints of the Most High'. Later apocalyptic literature (the Book of Enoch, of disputed date in relation to the Christian era) took up this figure and developed it. In contrast to this interpretation, Geza Vermes believes that *bar nash(a)* was commonly used in first-century Aramaic as a kind of modest circumlocution referring to oneself,[26] rather like the English upper-class usage, 'one likes a gin and tonic before dinner'. Once again it is a disputed scholarly point how well this practice has been established outside the gospels.

The second consideration relates to the way in which the phrase is actually employed. Sometimes it is the equivalent of 'man' (the sabbath saying of Mark 2.28, par.); sometimes it seems to carry Vermes' circumlocutory sense (Matt. 16.13, cf. the parallels in Mark 8.27 and Luke 9.18, where 'Son of man' is replaced by 'I'; see also Matt. 8.20, par.). Sometimes it carries a clear apocalyptic reference and it appears to be associated closely with Jesus (the passion predictions (Mark 8.31, par.; etc.); the reply to the high priest in Mark 14.62, par.). At other times the overtones of Daniel are less clear and the connection with Jesus more problematic. 'For whoever is ashamed of me and of my words in this adulterous and sinful generation, of him will the Son of man also be ashamed, when he comes in the glory of his Father with the holy angels' (Mark 8.38, par.). Here there is an intimate connection between the attitude to Jesus and the attitude of the Son of man, but it is not at all clear that the two figures are identified.

Thirdly, all four gospels frequently attribute the phrase to Jesus, but in only one trivial case (John 12.34) is it ever on the lips of anyone else. Outside the gospels, it is used only once, by Stephen in Acts 7.56, in relation to a heavenly vision, and twice without the article in Revelation (1.13; 14.14), where images of Daniel 7 are clearly being used in connection with visionary experience.

Such are the facts. What are we to make of them? One suggestion is that 'the Son of man' was a title for Jesus employed by the very early Church, after its having been culled from Daniel 7, and that it was then read back into his earthly life when the gospels were composed. The lack of any attestation of that primitive usage makes it very unlikely, in my judgement, that the phrase is a Jewish-Christian invention. If it came from the post-Easter community, it is incomprehensible that a verse like Mark 8.38 should exhibit any ambiguity about the identification of Jesus with the Son of man. The persistence of the phrase in the gospel tradition (from Mark to John, if that is the chronological order of their writing), and its virtual absence elsewhere, seem much more probably explained by a correctly remembered attribution to Jesus himself. I am persuaded by Moule's arguments to see a significant

[26] Vermes (1983), pp. 163–8.

reference to Daniel 7, so that the Son of man was for Jesus 'a symbol of a vocation to be utterly loyal, even to death, in the confidence of ultimate vindication in the heavenly court'.[27]

What then of that unclear conjunction hinted at in Mark 8.38, between the heavenly figure and Jesus himself? Here one must venture down the way, persistently placarded with warning signs by cautious scholars, of attempting some estimate of Jesus' self-understanding. The reason for the care needed is that this is an area where one is most vulnerable to the presence in the sources of the retrospective effects of later Christian hindsight. Nevertheless, the attempt must be made if we are to succeed in our aim of finding a credible picture of the one who was the initiator of the Christian movement. 'Nothing comes of nothing', and so much has come of Jesus that there must have been something present in him in the first place which was commensurate with the effect it produced. I think New Testament theologians in their attempts to reconstruct Christian origins need something of the corrigible boldness displayed by cosmologists in their approach to a similar task.

There are two extremes to be avoided. One is to attribute to Jesus such extraordinary powers that he effectively ceases to be credibly a recognizable human being. The other is so to recoil from this error that one treats him as if he were an uninteresting mediocrity. A good test case is provided by the so-called 'passion predictions', whose status is important in relation to an assessment of Jesus' self-understanding and purpose. Three times in each synoptic gospel, Jesus is portrayed as telling his disciples that he will be rejected and killed and after three days rise again (Mark 8.31; 9.31; 10.33–4; and par.). Many scholars regard these passages as prophecies-after-the-event, inserted by the post-Easter Church. I do not doubt that their form has been influenced by retrospection. I cannot think that Jesus saw his future laid out before him in fine detail, for I believe he lived a truly human life to which such precise foreknowledge would be foreign. But equally I cannot believe that he did not see in general terms that rejection and execution awaited him in Jerusalem, or that he did not trust that nevertheless he would be vindicated. To believe less than that is to make him out to be lacking in insight and faith. The woodenness of some New Testament scholarship is apparent when it is suggested that there is some tension between what I have just said and the story of the agony in Gethsemane (Mark 14.32–42, par.), as if foresight dissolves the intensity and demand of the actual moment. Equally lacking in imagination is the suggestion that the story of the Garden must be made up because the sleeping disciples could not have known what was going on, as if one could not conceive of a restless panicky dozing – half flight

[27] Moule (1977), p. 14.

from awful reality, half inescapable consciousness of it. This deeply moving but conventionally unheroic story (compare it with the death of Socrates or the deaths of many subsequent Christian martyrs) must, I believe, be a true reminiscence and of some profound significance. It is also attested elsewhere, outside the gospel tradition (Heb. 5.7).

The basis of Jesus' understanding of his mission lay in his firm confidence in God his Father, not in a detailed foreknowledge of what would happen. I believe that this view is consistent with Mark 8.38's assertion of the central role of Jesus himself, while leaving a certain openness as to the form of the 'ultimate vindication in the heavenly court'. The point of view I am proposing envisages that Jesus was clear as to his having received a unique call from God (the story of the baptism by John is important here), but it allows him to find the nature of its fulfilment through its unfolding realization (a process pictured in the gospels as beginning with the temptations in the wilderness, immediately subsequent to the baptism). Such a notion of an evolving vocation is not foreign to the New Testament. One thinks of Luke (2.52) and of the writer to the Hebrews, who combines a high Christology (Heb. 1.1–4) with a frank acknowledgement of development (Heb. 5.8–10).

If this picture has elements of truth in it, one might expect to find that the evocative openness of interpretation characteristic of the parables, and of that polyvalent phrase 'the Son of man', might be found elsewhere in the utterances of Jesus, as a means of indicating an envelope of understanding within which his nature and destiny could be contained and explored. This encourages me to think that the gospel language about Jesus which is reminiscent of Old Testament concepts of God's Wisdom (Prov. 8.22–31, etc.) – language which we find in passages like Matthew 11.28–30 ('Come to me all who labour and are heavy laden . . .') and, most strikingly, in Luke 11.49 ('the Wisdom of God said, "I will send them prophets and apostles . . ." '), where in the parallel passage in Matthew (23.34) 'Wisdom' is replaced by 'I' – while perhaps moulded by later Christian thought, has an anchorage in the words and understanding of Jesus himself. Jeremias has argued from a consideration of the probable underlying Aramaic[28] that the 'Johannine thunderbolt from a clear synoptic sky' of Matthew 11.27, par.: 'no one knows the Son except the Father, and no one knows the Father except the Son and any one to whom the Son chooses to reveal him', is really a pithy parable (only a father really knows a son and a son his father) rather than an absolute Christological assertion. It seems to me that the choice of interpretation may not be that stark and clear cut, for this might be a case where the creative ambiguity of a parabolic saying was being used in an

[28] Jeremias (1971), pp. 56–61.

exercise of exploration, and that the words are neither simply general nor specifically assertive, but rather an allusive hint of what might be. I find the same heuristic possibility in the 'ransom' saying of Mark 10.45. Jesus is portrayed as one who demands a committed response, centred on himself (Matt. 8.21–2, par.; 10.37–9, par.). In this way one can see the conceivability (I believe, the likelihood) that there are present in the synoptic gospels seminal sayings of the historical Jesus, which in the light of post-Easter reflection could lead to the Christ of the Johannine discourses. What is being suggested is that Jesus' self-understanding was consistent with later incarnational reflection, without being the same as it. It is also being suggested that Jesus was not curiously unreflective about himself, but that he sought images of adequate self-understanding and that he expressed them in ways which were remembered. We must return to these issues in chapter 7, but if elements both human and divine meet in Christ, the preservation of the true humanity surely must mean that the historical person did not go around thinking of himself as God. Something altogether more nuanced than that would have been necessary. Even in John we find this is recognized, for there are both claims of strong identification ('I and the Father are one', John 10.30) and of clear distinction ('for the Father is greater than I', John 14.28). In all the gospels, Jesus is portrayed as one who needed to pray. James Dunn summarizes the conclusion of his own careful survey of these questions by saying:

> We cannot claim that Jesus believed himself to be the incarnate Son of God; but we can claim that the teaching to that effect as it came to expression in the later first-century Christian thought was, in the light of the whole Christ-event, an appropriate reflection on and elaboration of Jesus' own sense of sonship and eschatological mission.[29]

Concern with these historical questions is an inescapable task for a bottom-up thinker. He cannot accept Kierkegaard's celebrated assertion that 'If the contemporary generation had left nothing behind them but these words: "We have believed that in such and such a year God has appeared among us in the humble figure of a servant, that he lived and taught in our community and finally died", it would be more than enough.'[30] Without further evidence, how could one know that this was even conceivably true? Rather, I must ask with Leonard Hodgson (in his Gifford Lectures for 1955–7), 'What must the truth have been if it appeared like this to men who thought like that?'[31] While I shall want to go on in a

[29] Dunn (1980), p. 254.
[30] Quoted in Macquarrie (1990), p. 237.
[31] Quoted in Moule (1977), p. 6.

later chapter to acknowledge the importance of the Church's testimony to its Lord, I cannot accept a primacy of the preached Christ over the historical Jesus, of the kerygma over history, but rather I feel impelled to strive for a mutually consistent understanding of them both. No doubt, in our encounter with Jesus there is recognition as well as surprise, so that, in Macquarrie's words, 'We recognize the historical Christ as revelation because we already have in our constitution as human beings an ideal archetype which, we believe, we see fulfilled in him.'[32] Yet that seems only half the story, for Jesus would not seem at times so strange and mysterious if only recognition were involved. Because I believe (and will defend the belief in what follows) that we see in Jesus much more than an encouraging or illuminating exemplar, so that he opens up the availability of a relationship between God and humanity previously undreamed of, I could not for a minute go on to contemplate with Macquarrie the possibility that 'there may come a time when only the pure archetype will remain as the focus of a wholly rational religion, and the last links with the historical Jesus and with historical Christianity will have been snapped'.[33] In my view, Christianity will never become all 'top', with no historical 'bottom'.

So far we have been concentrating on the words of Jesus, but his deeds are also of importance. If there was an early sayings source Q, it may well be significant that it did not survive on its own, but became incorporated into the narrative action of Matthew and Luke. The sayings source which we do have, the second-century apocryphal Gospel of Thomas, comes from a gnostic environment, where there would have been a 'top' concern with enlightened knowledge and a recoil from a 'bottom' engagement with the specificities of history.

One of Sanders' 'almost indisputable facts' is that Jesus was a healer. That would not have set him apart in the ancient world, where claims of such happenings were quite widespread, though mainly associated with cultic sites rather than individuals. The dispute about healing on the sabbath (Mark 3.1–6, par.) requires that such healings actually took place, and in general the gospel stories of healing are so numerous and interwoven into the accounts that they cannot be excised in any natural or credible way. There are a variety of ways in which to think about Jesus' mighty works. We may suppose that the psychosomatic consequence of encounter with a charismatic personality, say bringing release from hysterical paralysis, played a part in these happenings. It is possible to think of some of the nature miracles (e.g., the stilling of the storm, Mark 4.35–41, par.) as being examples of profoundly significant coincidences, such as happen from time to time.

[32] Macquarrie (1990), p. 183.
[33] ibid., p. 185.

Other miracle stories resist this naturalizing tendency. It seems lame to explain away the feedings of the multitude (Mark 6.32–44, par.; Mark 8.1–10, par.) by saying that an example of unselfishness induced the many to bring forth the rations they had prudently retained for their private use. Changing water into wine (John 2.1–11) is an irreducibly unnatural happening. A difficulty is that these latter stories also carry an obviously high symbolic value as signifying the difference that the presence of Jesus makes, and so they can be conceived of as tales in the tradition composed for this purpose.

Different people will judge the matter in different ways. It is impressive how matter-of-fact the gospel accounts of miracles are. Although we are often told things such as that the crowds were 'amazed and glorified God, saying, "We never saw anything like this!" ' (Mark 2.12), there is little attempt to pander to a taste for the marvellous, in striking contrast to the concoctions of pious fancy found in the second-century apocryphal gospels. Only very occasionally (perhaps the coin in the fish's mouth, Matt. 17.24–7) do we find a tale of what seems mere wonder-working. Jesus' healings were not coercive interventions; they required co-operation (Mark 6.5–6, par.). Anthony Harvey has given a careful discussion in which he compares the miracle stories associated with Jesus with those current about other figures in the ancient world. He comes to 'the remarkable conclusion that the miraculous activity of Jesus conforms to no known pattern'.[34]

A judgement on miracles cannot be separated from a judgement on Jesus; there is an inescapable circularity. The more we have reason to think him exceptional, the more coherent is the possibility that he exercised exceptional powers. C. H. Dodd, who has a rather reserved attitude to the miracle stories, says that while such things do not happen in ordinary circumstances, 'the whole point of the gospels is that circumstances were far from ordinary. They were incidental to a quite peculiar situation, unprecedented and unrepeatable.'[35] It is conceivable that unexpected events occur in unprecedented circumstances.[36] The miracle stories cannot simply be dismissed because of an a priori certainty that such things 'cannot happen', but equally the circularity we have noted in their evaluation means that they can no longer be used simply as 'proofs' of, say, Jesus' divinity, in a way that pre-critical generations attempted (cf. Mark 8.11–12, par.).

While the gospels contain the account of Jesus' actions, they lead to the story of the Passion; the verbs change from the active to the passive voice.[37]

[34] Harvey (1982), p. 113.
[35] Dodd (1971), p. 44.
[36] Polkinghorne (1989a), ch. 4.
[37] cf. Vanstone (1982).

The most striking difference between the canonical gospels and the Gospel of Thomas is that the latter shows no concern with the death of Jesus. In contrast, Mark devotes six of his sixteen chapters to the last week of Jesus' life. In a sense, the preceding chapters are there to establish the nature of the one who was crucified. The next chapter must be devoted to the death of Jesus and to its aftermath.

6

Crucifixion and Resurrection

'For our sakes he was crucified under Pontius Pilate;
he suffered death and was buried.
On the third day he rose again . . .'

The most certain thing about Jesus is that he was crucified during the procuratorship of Pontius Pilate. This is even testified in Roman secular history, for Tacitus in his *Annals* (early second century), speaking of the Christians as 'a class hated for their abominations', says that the name derives from 'Christus . . . [who] suffered the extreme penalty during the reign of Tiberius at the hands of one of our procurators, Pontius Pilate'.[1] The Jewish historian Josephus, in his *Antiquities*, written somewhat earlier than the *Annals*, makes a similar reference, but in a passage which has certainly been tampered with in some respects by later Christian editors.

Crucifixion was not only a very painful and lingering death; it was also a shameful death, of a kind inflicted on slaves and felons. To the Jew it was a sign of God's rejection, because of the word in the *Torah* forbidding a hanged man to be left on a tree: 'for a hanged man is accursed by God' (Deut. 21.23; cf. Gal. 3.13). Martin Hengel tells us that he can only find a solitary reference to a crucified martyr in the whole of the rabbinic sources.[2] He suggests that the emphatic rejection of the blasphemous paradox of a crucified Messiah may have been the spur to the young Saul's persecuting zeal. Nor would such an idea have been more acceptable in the Gentile world. The notion of apotheosis through a noble death would have been familiar enough (for example, Achilles), but that is poles apart from this squalid and ignominious end, which would have seemed 'aesthetically and ethically repulsive to them'.[3] No wonder that the converted Paul said that Christ crucified was 'a stumbling block to Jews and folly to Gentiles' (1 Cor. 1.23). Speaking of the word *stauros* (cross), Hengel says, 'Only those who

---

[1] Barrett (1956), pp. 15–16.
[2] Hengel (1981), p. 44.
[3] ibid., p. 31.

understand how extremely offensive this word will have been to both Jewish and Gentile ears' will be able to grasp the impact of the repetition of 'cross' and 'crucified' in Mark 15.[4] For us the cross is an ecclesiastical symbol; in the ancient world it was a reminder of the torturing gallows.

A consideration of the particular circumstances of Jesus' death can only intensify these feelings. I have already spoken of Gethsemane (pp. 100–1), one of the most profound and human episodes in the gospels, with both its frank acknowledgement of Jesus' prayerful dread at the ordeal which faced him and also its testimony to his openness to the will of the Father. We learn something here of what it means to trust in God within the uncertainties of human history. Jon Sobrino comments concerning Gethsemane that 'non-knowing is an essential feature of Jesus' prayer . . . it becomes part of a deeper knowledge of the Father'.[5] This sombre scene is the prelude to the passion. Jesus is then deserted by his friends. Who can doubt the historicity of the story of Peter's denial, recorded in all four gospels (Mark 14.66–72, par.) and making a shaming reference to the outstanding leader of the early Christian movement? From the darkness of 'the place which is called The Skull' (Luke 23.33, par.) comes the cry of dereliction. 'My God, my God, why hast thou forsaken me?' (Mark 15.34, par.). Jesus' life appears to end in total failure. Yet

> No human death has influenced and shaped the world of late antiquity, and indeed the history of mankind as a whole down to the present day, more than that of the Galilean craftsman and itinerant preacher who was crucified before the gates of Jerusalem in AD 30 as a rebel and messianic pretender.[6]

The other great founders of religious traditions – Moses, the Buddha, Muhammad – die in honoured old age. Jesus' near-contemporary, the wonder-worker Apollonios of Tyana, with whom he is sometimes compared, is said to have faced voluntarily the prospect of martyrdom under Domitian but then to have been spirited away magically in the course of his trial. It was not fitting that a 'divine man' should be executed. Jesus is killed in middle life, apparently rejected by God and man. The one who spoke with authority ends his life in utter powerlessness. It seems the story of complete defeat.

The early Christian tradition speaks quite otherwise. Hengel describes the formula 'Christ died for us' and its variations (Rom. 5.8; 1 Cor. 15.3, etc.) as being, together with the assertion of the resurrection of Jesus, 'the most

[4] ibid., p. 43.
[5] Quoted in Gorringe (1991), pp. 89–90.
[6] Hengel (1981), p. 1.

frequent and most important confessional statement in the Pauline epistles and at the same time in the primitive Christian tradition in the Greek language which underlies them'.[7] The cross of Christ was seen not as defeat but as victory, a victory by which humanity had been reconciled to God (2 Cor. 5.14–21), even while it was still at enmity with him (Rom. 5.6–11).

The clue to the significance of Jesus lies in the mystery of his death. In that lonely and deserted figure hanging there on the tree, what do we see? A good man, like so many good men before and after him, finally caught and destroyed by the system? A man of megalomaniac pretensions, who eventually got the come-uppance he deserved? The Saviour of the world? Only God can answer such a question. It was the belief of the early Christians that he had done so by raising Jesus from the dead that first Easter day. It is no accident that Hengel associates 'God raised Jesus from the dead' with 'Christ died for us' as the twin formulae of the primitive Christian confession.[8]

The resurrection is the pivot on which Christian faith turns, and it is the principal purpose of this chapter to explore whether the belief that God raised Jesus from the dead is one that is credible for us today. Two kinds of thought are involved in attempting such an assessment. One is the movement from below which looks for historical evidence which might be held to motivate belief in so singular an occurrence. However successful that quest might be, by itself it would not be sufficient, for how we actually weigh that evidence will be influenced by the extent to which we can make sense of the notion of the resurrection of Jesus within our general world-view. On its own terms, the resolute scepticism of a David Hume, so memorably expressed in the well-known passages on miracles in *An Inquiry Concerning Human Understanding*, could never be overcome, whatever the quantity of circumstantial evidence available, because he has an unshakeable certainty that 'A miracle is a violation of the laws of nature; and as a firm and unalterable experience has established these laws, the proof against miracle, from the very nature of the fact, is as entire as any argument from experience can possibly be imagined.' Yet that confidence that the laws of nature were known with a certainty that extends even into realms of unprecedented and hitherto unexplored phenomena is one that was certainly falsified by the history of science subsequent to the eighteenth century, and it could never be pressed to dispose of an event like the resurrection of Jesus, which claims to be a particular act of God in a unique circumstance. If Jesus was just an ordinary wandering preacher, the chances are that, like all other men, he stayed dead; if he is more than that, then it is a coherent possibility that the

[7] ibid., p. 37.
[8] ibid., p. 70.

aftermath of his death revealed new phenomena. Equally, if God raised him from the dead, that is surely a sign that he was indeed more than a wandering preacher. We cannot escape from the circularity which we noted in relation to all intellectual inquiry in chapter 2, and in relation to the miracles of Jesus himself in chapter 5. Neither can we escape from the insistent problem of how it came about that a man, living in a peripheral province of the Roman Empire, leaving virtually no trace in contemporary secular history, writing no book, dying an ignominious death, nevertheless has been dominant in human life and thought ever since. 'Who is Jesus?' and 'Did God raise him?' are questions which ineluctably interact. However, we must begin somewhere, and, in my bottom-up fashion, it is with questions of evidence that I shall first be concerned.

It is absolutely clear that something happened between Good Friday and Pentecost. The demoralization of the disciples, caused by the arrest and execution of their Master, is undeniable. Equally undeniable is the fact that within a short space of time, those same disciples were defying the authorities who had previously seemed so threatening, and that they were proclaiming the one who had died disgraced and forsaken, as being both Lord and Christ (God's chosen and anointed one). So great a transformation calls for a commensurate cause. From the nineteenth century onwards, in the thought of people such as Renan and Bultmann, it has been suggested that what happened was a faith event in the minds of the disciples, a conviction achieved after a period of reflection, that the cause of Jesus continued beyond his death. Somewhat similar is the opinion of Edward Schillebeeckx, for whom the primary terms of early Christian belief about the fate of Jesus centre on his being exalted to life with God, a conviction held to arise from pondering the idea of the vindication of the righteous set out in passages such as Wisdom 2.17–3.4, and finding only secondary expression in the idea of resurrection. 'It was only when people began to see that the deliverance of Jesus was also a conquest of death itself . . . that the idea of resurrection forced itself upon all the early Christian communities everywhere as the best way of articulating the fact of Jesus' being "alive to God".'[9] Frankly that seems to me to be a wholly unconvincing interpretation of New Testament attitudes, as subsequent analysis will seek to show. The writing with the most sustained emphasis on the theme of exaltation is the Epistle to the Hebrews, in which the language of resurrection is not used explicitly, but it is surely implicit throughout that the one who 'sat down at the right hand of God' (Heb. 10.12) was the one whom God had 'brought again from the dead' (Heb. 13.20). The notion of the exaltation of the crucified was too paradoxical a concept to have arisen

[9] Schillebeeckx (1974), p. 538.

109

in first-century Palestine simply through a process of reflection. I agree with Macquarrie when he says he does not think that 'reading or remembering a passage of scripture which speaks in a general way of a hope for the righteous beyond death would have been nearly percussive enough to produce in Peter and the others the radical turn-around or conversion they had at that time'.[10]

Even less likely is the theory, first suggested in the nineteenth century, and still occasionally put forward, that Jesus swooned on the cross and revived in the cool of the tomb. In addition to the many historical implausibilities involved (the Romans knew how to execute), this idea fails to convince because, as David Strauss rightly emphasized, 'It is impossible that a being who had stolen half dead out of the sepulchre . . . could have given the disciples the impression he was a Conqueror over death.'[11] However, Strauss believed it was hallucinatory experience which could have been the trigger of the change in the disciples' attitude, an idea which goes back to Celsus in the second century. The remorse of Peter caused in him an abreaction after the trauma of denial, which then communicated itself to the fraught band of disciples in a psychological chain reaction; that is the way a modern person might express the thought. Such an explanation fails to account for the varieties of time and place associated in the tradition with the claimed appearances, including Paul's experience on the Damascus road which must have been three years or so after the crucifixion. It also makes a good number of unsubstantiated assumptions about the temperamental make-up of the disciples. Also, I must confess to an instinctive feeling that hallucinations, however vivid, could not have been the enduring basis of the vitality of the early Christian movement.

Sanders, in his role of cautious historian, prescinds from the issue of what went on:

> What is unquestionably unique about Jesus is the result of his life and work. They culminated in the resurrection and the foundation of a movement which endured. I have no special explanation or rationalization of the resurrection experiences of the disciples. Their vividness and importance are best seen in the letters of Paul . . . We have every reason to think that Jesus had led them to expect a dramatic event which would establish the kingdom. The death and resurrection caused them to adjust their expectation, but did not create a new one out of nothing.
>
> That is as far as I can go in looking for an explanation of the one thing which sets Christianity apart from other 'renewal movements'. The

[10] Macquarrie (1990), p. 312.
[11] Quoted in O'Collins (1987), p. 101.

disciples were prepared for *something*. What they received inspired them and empowered them. It is the *what* that is unique.[12]

I cannot rest content with that. It seems necessary and reasonable to go on to ask, What was the explanation offered by the disciples themselves?

The New Testament answer is that they believed that Jesus had been raised from the dead and that 'To them he presented himself alive after his passion by many proofs' (Acts 1.3). It is important to remember that the earliest account of the resurrection appearances does not occur in the gospels, but in Paul's first letter to the Corinthians, written in the middle fifties AD. He had founded the Corinthian church, and he reminds them that when he did so

> I delivered to you as of first importance what I also received, that Christ died for our sins in accordance with the scriptures, that he was buried, that he was raised on the third day in accordance with the scriptures, and that he appeared to Cephas, then to the twelve. Then he appeared to more than five hundred brethren at one time, most of whom are still alive, though some have fallen asleep. Then he appeared to James, then to all the apostles. Last of all, as to one untimely born, he appeared also to me. (1 Cor. 15.3–8)

When Paul says he delivered what he also received, it is reasonable to suppose that he is referring to teaching given immediately following his conversion, just a very few years after the crucifixion itself. Thus this testimony takes us back very close indeed to the events cited. The antiquity of the material is confirmed by the use of the Aramaic 'Cephas' for Peter, and by the reference to 'the twelve', a phrase which soon fell out of Christian usage. The style of the reference to the five hundred brethren makes it plain that an appeal to accessible evidence is being made, an appeal which Bultmann, with his maximal distrust of the historical, regarded as dangerous, but which I, of course, whole-heartedly welcome. It is entirely possible that this attestatory role is the reason why there is no reference in Paul's list to the witness of the women (prominent in the gospel accounts of appearances), since in the ancient male-dominated world their testimony would not have been validly acceptable.

The account in 1 Corinthians 15 is extremely spare, a simple list of witnesses. It concludes with Paul's own encounter with the risen Christ, referred to again by the apostle himself in Galatians (1.11–17), and three times described in Acts[13] (9.1–9; 22.6–11; 26.12–18; with minor variations

---

[12] Sanders (1985), p. 320.

[13] I take a higher view of the basic historical reliability of Acts than do some New Testament scholars.

between the accounts, which provide another example of the kind of degree of detailed consistency which even a single author thought it necessary to achieve in the first century). Pannenberg reminds us that Paul is the only writer whose words are certainly those of a resurrection witness.[14] One might suppose Paul's experience to be best categorized as a vision. Pannenberg uses that language, but he emphasizes that if that is so, vision is used in a distinctive sense, for elsewhere the New Testament proves perfectly capable of speaking of visionary experience in terms carrying much less weight of significance than is attached to the case of the resurrection appearances (e.g., Acts 23.11). 'If the term "vision" is to be used in connection with the Easter appearances, one must at the same time take into consideration that primitive Christianity itself apparently knew how to distinguish between ecstatic visionary experience and the fundamental encounters with the resurrected Lord.'[15] Paul clearly places much less emphasis on a remarkable 'vision and revelation of the Lord' which he had received (2 Cor. 12.1–7; plainly a coy self-reference) than on his Damascus road encounter, which is the ground of his apostleship. 'Am I not an apostle? Have I not seen Jesus our Lord?' (1 Cor. 9.1). It is critical for his authority that he should find a place in that list of witnesses, alongside Peter and James and the rest. So his experience must be comparable to theirs. To assess what that might be and to get beyond that spare enumeration, we have to turn to the appearance stories in the gospels.

Immediately one enters a strange, almost dream-like world, in which Jesus appears in rooms with locked doors and suddenly disappears again. There is considerable difference of account between the different gospels. This latter point is in marked contrast to the preceding stories of the passion. These certainly display variations of detail (particularly in relation to the trials which Jesus underwent), but perhaps to no greater degree than one might expect in traditions stemming from recollections of a confused and frightening twenty-four hours. They are plainly recounting the same broad sequence of events. The gospel treatments of the resurrection appearances are much more diverse.

Mark, at least as far as the authentic text available to us is concerned, does not give a description of any appearance of the risen Jesus, though one is foreshadowed in 16.7: 'he is going before you to Galilee; there you will see him, as he told you'. Scholars have argued whether there is a lost conclusion to the gospel which would have supplied the present lack. Our text ends with the words about the women at the tomb, 'they said nothing to any one,

[14] Pannenberg (1968), p. 77. This implies certain reservations about the authorship of other New Testament writings (such as John and 1 Peter).
[15] ibid., p. 94.

112

for they were afraid' (Mark 16.8). Part of the discussion has been whether it was possible for a Greek text to conclude with *gar* (for). It now seems that this is conceivable. Whether such an ending is convincingly fitting is another matter. It suits a certain modern taste to end in mystery and fear, but I greatly doubt whether a first-century writer would have seen it that way. Certainly, by the second century it was felt appropriate to construct the additions to Mark which figure in some manuscripts and many of our translations. These incorporate ancient tradition, but they cannot be considered to constitute an independent witness in relation to the other gospels.

Matthew records a meeting of the risen Jesus with the women (Matt. 28.9–10), at which he tells them he will meet his 'brethren' in Galilee, and subsequently such a meeting is described (Matt. 28.16–20). The account of the latter includes an articulated trinitarian formula which must surely be a quite late development in the tradition.

In Luke 24 everything happens in Jerusalem on the first Easter day itself. Jesus does not meet with the women, but he journeys to Emmaus with two of the disciples and later he appears to the assembled eleven, finally parting from them after he has led them out to Bethany. There is also a brief reference to an appearance to Peter ('Simon', Luke 24.34; a verse that has about it something of the air of an oft-repeated credal statement). The same author's Acts speaks in general terms of appearances stretching over a period of forty days.

The most extensive sequence of resurrection appearances is described in John. Jesus is seen by Mary Magdalene (John 20.11–18). Then he appears to the eleven, less Thomas, on Easter evening, and a week later again to them all, including Thomas this time, who utters the most unequivocal assertion of Jesus' divinity found in the New Testament: 'My Lord and my God!' (John 20.19–29). In what appears to be an appendix added to the gospel (compare 20.30–1 with 21.24–5), we are given the detailed story of an appearance by the lakeside in Galilee (John 21.1–23), which has some elements in common with an incident which Luke locates in Jesus' lifetime (Luke 5.3–7).

It is a somewhat confusing mass of material. Pannenberg believes that 'The Easter appearances are not to be explained from the Easter faith of the disciples; rather, conversely, the Easter faith of the disciples is to be explained from the appearances.'[16] However, his estimate of the gospel material is that

> The appearances reported in the Gospels, which are not mentioned by Paul, have such a strongly legendary character that one can scarcely find a historical kernel of their own in them. Even the Gospels' reports that

[16] ibid., p. 96.

correspond to Paul's statements are heavily coloured by legendary elements, particularly by the tendency toward underlining the corporeality of the appearances.[17]

I agree with the judgement about appearances leading to faith, but not with the assignment of almost all detail to the category of legend. Amid the variety of the appearance stories there is one element which is both unexpected and persistent. It is that there was difficulty in recognizing the risen Jesus. Dodd is right to say that the stories 'are all centred in a moment of recognition',[18] and to use that as an argument against their being assimilated to a category of vague mystical experience, but that moment of recognition is not easily attained. Mary Magdalene mistakes Jesus for the gardener; on the Lake of Galilee, only the beloved disciple has the insight to recognize that the figure on the shore is the Lord; the couple walking to Emmaus only realize who their companion has been at the moment of the breaking of bread and his disappearance; with great frankness Matthew tells us that when Jesus appeared on the mountain in Galilee 'they worshipped him; but some doubted [*edistasan*]' (Matt. 28.17). This would be a strange motif to recur in stories which were merely made up. It seems likely to me that, on the contrary, it is the kernel of a genuine historical reminiscence.

The corporeality claimed for the risen Jesus is emphasized in Luke, where Jesus encourages the disciples to handle him, 'for a spirit has not flesh and bones as you see that I have' (Luke 24.39), and where he eats some broiled fish; and to a lesser extent in John where, though Mary Magdalene is told not to cling to him, the disciples are invited to inspect the wounds, and where, by implication, Jesus might be thought to have joined in the lakeside meal.[19] What one makes of this – whether one automatically reaches for the category 'legendary' – depends upon one's understanding of human embodiment and its conceivable destiny. (I shall return to that issue in a later chapter.) Commenting on the Lukan passage, G. B. Caird says that 'to a Jew a disembodied spirit could only seem a ghost, not a living being, but a thin, unsubstantial carbon-copy which had somehow escaped from the filing-system of death'.[20] J. I. H. McDonald says that 'Luke was ruling out any truck with notions such as subjective vision, psychic peculiarity or insubstantial shade to account for the risen figure of Christ. Jesus is real and is found among the living.'[21] I am very wary of those who want to take too

[17] ibid., p. 89.
[18] Dodd (1971), p. 40.
[19] One might ask the question, Did the risen Christ breathe? That too would involve exchange between his glorious body and the environment.
[20] Caird (1963), p. 261.
[21] McDonald (1989), p. 107.

exclusively spiritual a view of anything relating to humanity. For that reason I do not warm to those who use abstraction from the Acts accounts of the Damascus road experience to argue that luminous glory, rather than a focus on personal identity, lies behind the appearance stories. This led J. M. Robinson to claim that the vision of the exalted Christ experienced by the seer of Patmos (Rev. 1.9–20) is 'the only resurrection appearance in the New Testament that is described in any detail'.[22] Commentators have sometimes suggested, in a similar vein, that the story of the transfiguration (Mark 9.2–8, par.) is a displaced resurrection appearance. These seem to me to be implausible manipulations of New Testament material, whose authors should be given greater credit for knowing what they were doing. As J. A. Baker points out, the actual appearance stories in the gospels conform to neither of the contemporary models for post-mortem phenomena, which are a dazzling heavenly figure or a resuscitated corpse.[23]

Acknowledgement of a degree of corporeality in the appearance accounts is far from equating the resurrection with a mere resuscitation. Whatever we may make of the stories of those whom Jesus restored to life (Mark 5.21–43, par.; Luke 7.11–17; John 11.1–44), there is no question that they were destined eventually to die again. They were resuscitated, not resurrected. Jesus, however, is raised to endless life; his resurrection body is transmuted and glorified, possessing the unprecedented properties that allow him to appear and disappear in locked rooms, yet bearing still the scars of the passion. What such continuity and discontinuity might mean is tentatively explored by Paul in 1 Corinthians 15 (vv. 35–50) in terms of our own eventual destiny beyond death. He warns his readers against a resuscitatory reductionism ('flesh and blood cannot inherit the kingdom of God, nor does the perishable inherit the imperishable', v. 50). Yet the tone of the passage is also against a spiritual reductionism, for it is the resurrection *body* which is being discussed. We must return to this in chapter 9, but now it is necessary to consider a second line of evidence which has a potential bearing upon such questions. I refer, of course, to the stories of the empty tomb.

All four gospels contain accounts of how, once the sabbath was over, women came to the tomb to attend to the body of Jesus, only to find the stone rolled away and the tomb empty (Mark 16.1–8, par.). The stories differ in details of timing in relation to dawn, the names of the women involved and the number of angelic messengers they encountered. However much these discrepancies may have disturbed Edward Pontifex in the atmosphere of narrow literalism portrayed in *The Way of All Flesh*, they are unlikely to

[22] Quoted in O'Collins (1987), p. 212.
[23] Baker (1970), pp. 253–5.

disconcert us. We can accept such variation without believing that this by itself casts doubt on the core tradition.

For some, these stories are the strongest evidence for the resurrection. Why did the Jerusalem authorities not nip the nascent Christian movement in the bud by exhibiting the mouldering body of its leader? It is incredible to suggest that the disciples stole the body in an act of contrived deceit, and unbelievably lame to suggest that the women went to the wrong tomb, so that it all arose from a mistake. The only credible reason for the emptiness of the sepulchre was that Jesus had actually risen. So the argument goes.

A somewhat more careful assessment is required. The first explicit account of the empty tomb is in Mark, written some thirty-five years or so after the event. It is suggested by some scholars that we have here a second-generation story, made up as the expression of an already existing conviction (perhaps based on the appearances) that Jesus had survived death. Even the fact of a separate tomb at all is held to be questionable, for it was the common Roman practice to inter executed felons in the anonymity of a common grave. A number of points may be made in response.

While it is notorious that Paul does not refer explicitly to the empty tomb in his extant letters, not only is the argument from silence particularly dangerous when applied to such occasional writings, but also the occurrence of the phrase 'was buried' in that extraordinarily spare summary in 1 Corinthians 15 seems clearly to indicate that a special significance attached to the burial of Jesus. It seems very hard to believe that a Jew like Paul, whose background of thought would have been one emphasizing the psychosomatic unity of the human being, could have believed that Jesus was alive but that his tomb still contained his mouldering body. James Dunn concludes a survey of first-century Pharisaic thought and practice by saying, 'the ideas of resurrection and of empty tomb would naturally go together for many people. But this also means that any assertion that Jesus had been raised would be unlikely to cut much ice *unless his tomb was empty*.'[24]

There is archaeological evidence from Palestine later in the first century which shows that a crucified man was, in that case, buried separately and not assigned to a common grave. Thus the story of Jesus' separate burial is not impossible. If it were a made-up story, it is hard to see why Joseph of Arimathea and Nicodemus are the names associated with it, since these figures do not play any prominent part in the subsequent story of the Christian movement. The most natural explanation of their assignment to an honoured role is that they fulfilled it.

Equally, if the discovery of the empty tomb were a concocted fiction, why, in the male-dominated world of that time, were women chosen to play

[24] Dunn (1985), p. 67.

the key parts? Far and away the most natural answer is that they actually did so. Of course, there are oddities about the story. How did the women imagine they were going to cope with the heavy stone blocking the entrance? (This problem is acknowledged in the account: Mark 16.3.) After three days, in that hot climate, would it not have been too late to attend to the corpse? However, contemporary understanding held that corruption set in on the fourth day (cf. John 11.17, 39). John alone suggests that some preliminary precautions had been taken on the Friday evening (John 19.39–40). Such problems are, perhaps, more characteristic of the roughness of reminiscence than the smoothness of composition and, in any case, one should not expect coolly logical behaviour from women still distraught at the execution of their revered Master.

Whatever difficulties twentieth-century scholars may feel about the empty tomb stories, they do not seem to have been shared by critics of Christianity in the ancient world. As a bitter polemical argument sprang up between Judaism and the Church, it was always accepted that there was a tomb and that it was empty. The critical counter-suggestion was that the disciples had stolen the body in an act of deception, an explanation which I regard as incredible. Just how far back this argument can be traced is indicated by the story of the watch set on the tomb (Matt. 27.62–6; 28.11–15). I consider this to be a patently fabricated tale from a Christian source, concocted precisely to rebut the canard that the disciples had been grave-robbing. There is clear evidence, then, that in the first century those hostile to Christianity nevertheless accepted that the tomb had been found empty. A confirmatory consideration is the complete lack of any evidence of a cult associated with the burial place of Jesus. Ancient Jewish piety was much given to respectful veneration of the tombs of prophets and patriarchs (cf. Matt. 23.29). The total absence of this in the case of Jesus strongly suggests that from the first it was realized that for him the tomb was an irrelevancy. Christian interest in the possible burial place only dates from later centuries, when an increasing engagement of Christian thought with history led to giving attention to sites associated with Jesus' life.

Thus there are many reasons for taking seriously the tradition of the empty tomb, in addition to the tradition of the appearances of the risen Christ. Dodd summarizes his assessment of the gospel writers' narratives by saying, 'It looks as though they had a solid piece of tradition, which they were bound to report because it came down to them from the first witnesses, though it did not add much to the message they wished to convey, and they hardly knew what use to make of it.'[25] He is emphasizing the fact, contrary to some modern apologetic strategies, that the gospels do not

[25] Dodd (1971), p. 172.

present the empty tomb as a knock-down argument for the truth of the resurrection. Rather, it requires explanation. Hence the need for the message of the angel, 'He has risen, he is not here' (Mark 16.6, par.); the flow of understanding is from resurrection to absence of the body, rather than the reverse. In the story of Peter and the beloved disciple at the tomb (John 20.3–10), it is only the latter who has the insight to recognize unaided what has happened. For the others, the discovery of the emptiness of the tomb is, at first, disorientating; 'a nasty shock'.[26] There is no easy triumphalism in these stories, which itself makes one the more inclined to accept them as stemming from authentic recollection. From the point of view of the New Testament, it is the resurrection which explains the empty tomb rather than the empty tomb proving the resurrection.

There are twentieth-century Jewish writers who accept the emptiness of the tomb without thereby being driven to embrace Christianity. Geza Vermes concludes: 'In the end, when every argument has been considered and weighed, the only conclusion acceptable to the historian must be . . . that the women who set out to pay their last respects to Jesus found to their consternation, not a body, but an empty tomb.'[27] The orthodox Jew, Pinchas Lapide, goes further. He believes that Jesus was raised from the dead, but he does not accept him as the Messiah, let alone the incarnation of God. 'Thus, according to my opinion, the resurrection belongs to the category of the truly real and effective occurrences, for without a fact of history there is no act of true faith.'[28] I shall want to return in a later chapter to the question of the theological significance of the empty tomb.

In the meantime, let us add to the circumstantial discussion of the traditions of the appearances and of the empty tomb certain other inferential considerations which have a bearing on the resurrection. One of these is the existence of Sunday, the first day of the week, as the special Christian day, despite the Church's origin in a Jewish setting for which the sabbath, the seventh day, was the day of special religious significance. From the very earliest Christian times, Sunday has been the Lord's day. The obvious explanation is that this originated in the belief that it was the day of resurrection, the 'third day'[29] from Good Friday, mentioned in the primitive teaching (1 Cor. 15.4). Paul says this was so 'in accordance with the scriptures', and some have wondered whether the emphasis on the third day derives from Hosea 6.2, 'After two days he will revive us; on the third

[26] McDonald (1989), p. 140.
[27] Vermes (1983), p. 41.
[28] Lapide (1984), p. 92.
[29] Chronology in the ancient world was inclusive; one counted both the first and last days of a period.

day he will raise us up,' instead of its being a historical reminiscence. I would judge that it is much more likely that the first Christians knew about the third day and then sought out this rather obscure verse as a scriptural confirmation of it, rather than the other way round. I cannot believe that the ancient tradition of the Lord's day was set upon so slender an exegetical foundation as Hosea would provide.

Strictly, of course, the third day is preserved as the day on which the tomb was found empty and the Lord first appeared to his disciples. There is an impressive discretion in the gospels which restrains them from any attempt to describe God's act of resurrection itself. The well-known passage in the apocryphal Gospel of Peter shows us how inadequate pious fancy proved for this task when second-century curiosity encouraged the concoction of such descriptions.

The final inferential consideration is one which the non-believer may treat with some reserve, but which is unquestionably part of the Christian testimony. The Church in every century has characteristically spoken of Jesus as its living Lord in the present; it does not look back to him as a revered founder-figure of the past. In a metaphor drawn from cosmology, one could say that the Church is the 'background radiation' persisting after the primary event of the raising of Jesus. Its quintessential claim is: 'Jesus lives!' There is a striking contrast here with the other great religious traditions of the world, which do not speak of their revered founders in such contemporary terms.

There is also a striking contrast between the early Christian assertion that a recently known figure was resurrected within history and the contemporary expectations of the ancient world. Many, but not all (Mark 12.18–27, par.), Jews looked for a general resurrection of the dead at the end of time. In John Robinson's words, 'No one expected to find a grave empty in the middle of history.'[30] Hence the initial consternation at its discovery. The embarrassing scene in Matthew, where the death of Jesus provokes an earthquake and the saints come out of their tombs (Matt. 27.51–4), is precisely an attempt in pictorial language to associate with Jesus within history those eschatological events which really belong beyond history, in the effort to cope with the perplexity involved in his *historical* resurrection. Comparisons with Egyptian stories about Isis or Greek stories about Aesculapius miss the point that these refer to legendary figures of the mythological past, not to a wandering preacher who but yesterday was drawing the crowds.

I hope I have made it clear that there is motivation for the belief that Jesus was raised from the dead (the most ancient expression is always in the

---

[30] Robinson (1972), p. 132.

passive; it is a great act of God, not a final miracle of Jesus, which is being asserted). We now have to ask the question whether the motivation provided is in fact strong enough to support the extraordinary claim being made. Such an assessment will depend upon the second movement of thought about which I spoke when I introduced the question of the resurrection. Can it make sense within a general understanding of God and his ways with humanity that, alone of all who have ever lived, this man was restored to unending life in an act which, although it transcends history, nevertheless is embedded in history?

Inevitably we return to the hermeneutical circle in which the significance of Jesus and the truth of his resurrection inextricably interact with each other. The modern theological writer who has expressed this most clearly and emphatically is Wolfhart Pannenberg. He says bluntly, 'Jesus' unity with God was not yet established by the claim implied in his pre-Easter appearance, but only by his resurrection from the dead,'[31] and later he asserts that 'Apart from Jesus' resurrection, it would not be true that from the very beginning of his earthly way God was with this man.'[32] It is the resurrection which makes it plain who Jesus is. Paul says that 'if Christ has not been raised, then our preaching is in vain and your faith is in vain' (1 Cor. 15.14). A top-down thinker like Macquarrie demurs: 'I doubt very much whether in the case of such a complex system of beliefs as Christianity, such a simplistic mode of falsification is possible.'[33] We bottom-up thinkers view things differently, for we are open to the possibility of critical events on which an understanding pivots. For sure, once a complex physical theory, such as special relativity, has achieved 'well-winnowed' status, it will not be falsified by the first claim of an adverse experiment, but special relativity would never have come into being at all if Michelson and Morley had measured a non-zero velocity of the Earth through the aether in their famous experiment. I am not claiming that the whole of a developed traditional Christology can be read out from the resurrection, but I do believe that if we cannot make the claim 'Jesus lives', the ambiguity of his death remains an unresolved enigma and the significance of his life and message seem at most a brave gesture in a hostile world. Christianity would not have come into being without the resurrection of Jesus.

Pannenberg says that 'if the cross is the last thing we know about Jesus then – at least for Jewish judgement – he was a failure'.[34] Moltmann concurs: 'As a merely historical person he would long have been forgotten,

[31] Pannenberg (1968), p. 53.
[32] ibid., p. 341.
[33] Macquarrie (1990), p. 406.
[34] Pannenberg (1968), p. 112.

because his message had already been contradicted by his death on the cross.'[35] It seems to me entirely possible that if Jesus had not been raised from the dead we would never have heard of him. Lapide, from his Jewish perspective, speaks of 'the *must* of the resurrection': 'Jesus *must* rise in order that the God of Israel could continue to live as their heavenly Father in their hearts; in order that their lives would not become God-less and without meaning.'[36] He serves to remind us that the resurrection is not only the vindication of Jesus. It is also the vindication of God: that he did not abandon the one man who wholly trusted himself to him. Moreover, we begin to see here some glimmer of a divine response to the problem of evil. If Good Friday testifies to the reality of the power of evil, Easter Day shows that the last word lies with God. David Jenkins writes:

> We do not see how the purposes of love can be reconciled with the purposelessness of evil, but we do see that the human being who embodies the pattern of a loving God is both submerged in the destructiveness of evil and emerges from it as a distinctive, loving and personal activity. The Logos of the cosmos is not a mythological theory but a crucified man. The hope of personal sense and fulfilment lies neither in ignoring evil nor in explaining evil, but in the fact that Jesus Christ endured evil and emerged from evil.[37]

I shall return to this theme in the next chapter.

Finally, the resurrection of Jesus is the vindication of the hopes of humanity. We shall all die with our lives to a greater or lesser extent incomplete, unfulfilled, unhealed. Yet there is a profound and widespread human intuition that in the end all will be well. Max Horkheimer spoke of the wistful longing that the murderer should not triumph over his innocent victim. The resurrection of Jesus is the sign that such human hope is not delusory. It is the antidote to that human dread of the threat of non-being on which the existentialist tradition from Kierkegaard onwards has laid such emphasis. This is so because it is part of Christian understanding that what happened to Jesus within history is a foretaste and guarantee of what will await all of us beyond history. 'For as in Adam all die, so also in Christ shall all be made alive' (1 Cor. 15.22); 'he is the beginning, the first-born from the dead, that in everything he might be pre-eminent' (Col. 1.18).

The resurrection of Jesus is a great act of God, but its singularity is its timing, not its nature, for it is a historical anticipation of the eschatological destiny of the whole of humankind. The resurrection is the beginning of

[35] Moltmann (1974), p. 162.
[36] Lapide (1984), p. 89.
[37] Jenkins (1967), p. 89.

God's great act of redemptive transformation, the seed from which the new creation begins to grow (cf. 2 Cor. 5.17). Pannenberg says, 'Only the *eschaton* will ultimately disclose what really happened in Jesus' resurrection from the dead.'[38] Rahner says that 'By his resurrection and ascension Jesus did not merely enter into a pre-existent heaven; rather his resurrection created heaven for us.'[39] When John Robinson says that 'The "new thing" that God is doing is always concerned with the re-creation of the old rather than with its scrapping and supercession – He takes up the continuities to remake them,'[40] one thinks immediately of the continuity-in-discontinuity of the resurrected Jesus, the glorified body which bears the scars of the passion.

When Jesus himself was questioned by the Sadducees about the idea of a general resurrection at the end of time, he based his answer on the faithfulness of God, the God for whom Abraham, Isaac and Jacob were not simply persons who had served their turn and then were discarded, but were people of continuing significance to him (Mark 12.18–27, par.). John Baillie comments:

> The argument is unanswerable; and is indeed the only unanswerable argument for immortality that has ever been given, or ever can be given. It cannot be evaded except by a denial of the premisses. If the individual can commune with God, then he must matter to God; and if he matters to God, he must share God's eternity. For if God really rules, He cannot be conceived as scrapping what is precious in his sight.[41]

That, in a nutshell, is the case for the fittingness of a belief in the resurrection of Jesus as well as the belief in a destiny for ourselves in him.

When I first wrote about these matters, I concluded that when we consider the New Testament evidence,

> the only explanation which is commensurate with the phenomena is that Jesus rose from the dead in such a fashion (whatever that may be) that it is true to say that he is alive today, glorified and exalted but still continuously related in a mysterious but real way with the historical figure who lived and died in first-century Palestine.[42]

I stand by that judgement today.

---

[38] Pannenberg (1968), p. 397.
[39] Quoted in Macquarrie (1990), p. 410.
[40] Robinson (1972), p. 49.
[41] Quoted in McDonald (1989), p. 71.
[42] Polkinghorne (1983), p. 89.

'he ascended into heaven and is seated
at the right hand of the Father'

The language of the creed, and the New Testament passages on which it is based (Luke 24.51; Acts 1.6–11; 2.34–6; Heb. 8.1, etc.), is heavily symbolic at this point. We are not committed to the quaint picture, sometimes found in medieval stained glass, of the Lord's feet projecting from the underside of a cloud, as he sets out on his space-journey to the heavenly realm. In scripture a cloud is the symbol of the presence of God (Exod. 19.16; Dan. 7.13; Mark 9.7, par.), and its role in the story of the ascension is to emphasize the divine authority of the exalted Christ. A similar purpose is served by the mythological language of the heavenly session. The words of Psalm 110.1: 'The Lord says to my lord: "Sit at my right hand, till I make your enemies your footstool," ' afforded the early Church some clue to how the Lordship of Christ was related to the fundamental Lordship of God – a clue quite probably originating in the probing words of Jesus himself (Mark 12.35–7, par.) – and that made this verse the Old Testament text most frequently referred to by the writers of the New Testament.

The ascension and session symbolize two profound mysteries of Christian belief: that in Christ the life of humanity is taken up into the life of God himself, and that, though 'we do not yet see everything in subjection to him' (Heb. 2.8), yet Jesus Christ is the invincible agent of the ultimate purpose of God.

7

~~~~~

Son of God

'The only Son of God, eternally begotten of the Father,
God from God, Light from Light,
true God from true God, begotten, not made,
of one being with the Father'

When we turn from the gospels to the other writings of the New Testament, which overtly present a post-Easter account of the impact of Jesus, we enter a realm of discourse where the dominant impression is of people groping for concepts capable of doing justice to their experience. The evidence is of an event which cannot be contained within conventional limits of thought. Acts (which I believe preserves significant remembrance of very primitive Christian teaching) presents Peter as already proclaiming on the day of Pentecost that 'God has made him both Lord and Christ, this Jesus whom you crucified' (2.36).

The gospels portray Jesus as displaying some reserve about the title of Christ, which is not to be proclaimed abroad (Mark 8.27–30, par.), doubtless because it is a suffering Messiah, rather than a victorious warrior, who corresponds to Jesus' conception of his destiny (Mark 8.31–3, par.). Dunn suggests that we should interpret Jesus' enigmatic replies to Caiaphas and Pilate, when questioned about his messianic claims (Mark 14.62; 15.2, par.; Dunn takes a variant textual reading in Mark 14.62), as amounting to, 'If you want to put it that way.'[1] By the time of Paul's epistles, Christ has been so closely associated with Jesus that it functions as a kind of surname, rather than a title.

However paradoxical the idea of a crucified Messiah would have been to the first-century Jewish mind, the concept of God's chosen and anointed one is capable of being understood in purely human terms. The same is true originally of the title 'Son of God', for the Old Testament (e.g., Ps. 2.7) uses such language of the king, God's adopted viceregent. In the wider Hellenistic

---

[1] Dunn (1977), p. 42.

world, men with marvellous powers were often called 'sons of God'. That Jesus saw himself in some special relation of sonship is suggested not only by his use of *abba*, but also by the parable of the wicked husbandmen (Mark 12.1–11, par.; also Gospel of Thomas 65.66). I do not take the view that Jesus never used allegory and that therefore it follows that this story must be a construction of the early Church. Rather, the absence of a vindication of the son, and the appropriateness of the setting of the parable in those fateful days after the entry into Jerusalem, persuade me that it goes back to Jesus himself and expresses, in a challenging way, his understanding of his role as God's ultimate envoy ('a beloved son').[2]

By the time we get to Paul, sonship language has undergone a very considerable development. A passage such as Romans 8, where Paul talks of what 'God has done . . . sending his own Son' (v. 3), that he 'did not spare his own Son but gave him up for us all' (v. 32), so closely identifies Jesus with God's decisive action for the redemption of all humanity, and depends for its force upon a uniquely intimate connection between him and the Father, that we have moved far beyond what can be contained within representative notions of human kingship. The fourth gospel continues the process yet further (John 3.16–21; 5.19–29; etc.). It is impossible in that gospel not to spell Son with a capital 'S'. Dunn concludes his survey of the New Testament usage of the title Son of God by saying:

> Whether the thought focuses on Jesus' resurrection and parousia, or his anointing at Jordan, or on his birth, or embraces the whole of time, it is the language of divine sonship which appears again and again, sometimes without rival . . . The emergence of 'Son of God' as the dominant title for Christ in the fourth century was well justified by its importance in earliest Christology.[3]

I do not think these developments in first-century Christian thought represent the results of unbridled metaphysical speculation. Rather, we see the struggle to do justice to the encounter with this man, in his life and death and resurrection, which nevertheless simply cannot adequately be expressed in human terms alone. Perhaps the clearest indication of this is the application of the title 'Lord' to Jesus. Dunn reminds us that this occurs nearly 230 times in the Pauline epistles alone,[4] for which 'Jesus is Lord' is a central affirmation (Rom. 10.9; 1 Cor. 12.3; etc.). *Kyrios* is a loaded word, for in the Greek translation of the Old Testament it was used (just as LORD is

[2] cf. Dodd (1961), pp. 93–8; Jeremias (1972), pp. 70–7. I do not accept Jeremias' interpretation of the parable.
[3] Dunn (1980), p. 64.
[4] Dunn (1977), p. 50.

used in many English versions) as an utterable representation of the divine name (YHWH), which no pious Jew could ever speak aloud. Moule concludes from a careful survey of the implications of *Kyrios* (and its Hebrew and Aramaic equivalents, *adon* and *mar*) that its use for Jesus 'had behind it both Semitic and Greek associations of considerable significance'.[5] The fact that Paul can apply to Jesus Old Testament texts which in their original refer to God himself, can only reinforce this view. The great hymn of Philippians 2.6–11 speaks of the exalted Jesus at whose name 'every knee should bow . . . and every tongue confess that Jesus Christ is Lord, to the glory of God the Father' – an unmistakable echo of a sternly monotheistic passage in Isaiah (Isa. 45.23; see also Rom. 10.13; 1 Cor. 2.16 for similar adaptations of Old Testament passages; cf. Heb. 1.8–12). The fact that many scholars regard the Philippians passage as being a quotation of pre-Pauline material places this type of usage very early indeed in the development of Christianity. This judgement is confirmed by the use of the Aramaic formula *maranatha* ('Lord, come!') in a Greek text (1 Cor. 16.22).[6] Yet New Testament writers, however much their language about Jesus carries overtones of divinity, are extremely discreet and reserved about actually calling him 'God'. The only candidate example in Paul is Romans 9.5, where the sense depends upon a disputed point on how to read and punctuate the Greek text. The clearest New Testament statement of Jesus' unqualified divinity is Thomas' confession, 'My Lord and my God!' (John 20.28).

The New Testament writers raise the question of Jesus' relation to the divine without resolving it. This is set out most plainly in the opening formulae of many of the letters. Paul starts almost all his epistles with the greeting: 'Grace to you and peace from God our Father and the Lord Jesus Christ' (Rom. 1.7; 1 Cor. 1.3; etc.). It is a very strange sentence. God and Jesus are bracketed together, without any apparent feeling of incongruity. (How unthinkable it would have been for a Jew so to associate God and Moses.) God is the Lord and yet Jesus is Lord also, without the two being identified. Moule comments on these opening formulae that they are 'nothing short of astounding, when one considers that they are written by monotheistic Jews with reference to a figure of recently past history'.[7] They clearly result from a struggle to do justice both to the 'Christ-event' and to that fundamental Israelite assertion that 'The Lord our God is one Lord' (Deut. 6.4). Their intellectual instability is manifest. Further thought must lie ahead, grappling with how the Lordship of Christ and the Lordship of God are to be reconciled and understood.

---

[5] Moule (1977), p. 149.
[6] See the discussion in Cullmann (1963), ch. 7.
[7] Moule (1977), p. 150.

Lord and Christ are but two of the titles which constellate around Jesus in the attempt to find a fitting manner of speaking about him.[8] Others include the prophet foretold by Moses (Acts 3.22); the suffering servant (hinted at in the 'servant' (*pais*) of Acts 3.13, 26; 4.27, 30 (echoing the Septuagint of Isa. 53); see also 1 Pet. 2.24–5 – there seems to be an ancient Petrine association present in this tradition); the second Adam (explicit in Rom. 5.12–17; 1 Cor. 15.22, 45–9; implicit, many commentators think, in the language of passages like Phil. 2.6–11; so that Dunn emphasizes 'how important and widespread was Adam theology in earliest Christianity'[9]); divine Wisdom (1 Cor. 1.18–25), culminating in the Cosmic Christ of Colossians (1.15–20), in the picture of the one who 'reflects the glory of God and bears the very stamp of his nature, upholding the universe by his word of power' (Heb. 1.3), and in the Word (*Logos*) of the prologue to John's gospel (1.1–18). I have arranged these titles in an ascending Christological sequence, rising from that capable of expression in purely human terms, through the notion of a heavenly figure to that which demands the status of divine or quasi-divine. And all these descriptions, remember, focus on a historical figure, only recently dead. Moule comments that a

> most remarkable feature of the Christians' presentation of Jesus was that not only was Jesus claimed to be an anointed one, but also a whole welter of other figures converged on him – figures both for individual saviours and also for the realization of the true destiny of Israel as a whole.[10]

Dunn concludes a survey of the diversity of early confessional formulae by saying that

> the distinctive feature which comes to expression in all the confessions we have examined, the bedrock of the Christian faith confessed in the NT writings, is *the unity between the earthly Jesus and the exalted one who is somehow involved in or part of our encounter with God in the here and now.*[11]

We are so familiar with this piling of image and title on Jesus in the attempt adequately to do justice to what the Christ-event meant for those who first experienced it, that we often fail to recognize what an extraordinary phenomenon it is. There is a hazy feeling that the ancient world was so open to tales of theophanies that it is no great surprise that Jesus attracted such assessments. Yet the association of a historical man with

---

[8] See, for example, Dunn (1980); Cullmann (1963).
[9] Dunn (1980), p. 125.
[10] Moule (1977), p. 151.
[11] Dunn (1977), p. 57.

such depth of divine description appears to be unprecedented. Dunn's conclusion, after a search for parallels, is that

> we have found nothing in pre-Christian Judaism or the wider religions of the Hellenistic world which provides sufficient explanation of the origin of the doctrine of the incarnation, no way of speaking about God, the gods, or intermediary beings which so far as we can tell would have given birth to this doctrine apart from Christianity.[12]

People sometimes compare Jesus with Jewish charismatics like Honi the Circle-drawer, or wonder-working 'divine men' like Apollonios of Tyana, or even rabble-rousing insurgents like Theudas. Yet who are these people? Today known only to scholars, even in their own time they never came near to attracting to themselves such astonishing claims to be a focus of the divine and the fulfilment of God's saving purposes for all humanity. The New Testament itself contains an instructive comparator. The gospel stories of John the Baptist make it plain that he was a profoundly charismatic figure and from Acts (18.25; 19.1–5) it is clear that the movement associated with him persisted for some time. Whatever perplexity, or even anxiety, this may have given the early Christians, it is certain that no one was tempted to speak of John in terms other than the strictly human. Why is Jesus so different? I think the answer must be that God was perceived to be present in and with him in some unique, unprecedented way. Clearly, the resurrection was a spur to this judgement, but it was also motivated by the new life those early believers had found for themselves in Christ.

The first century is the age of fertile Christological exploration, in which titles and images condense upon the historical figure of the crucified Galilean teacher. Macquarrie has compared the rapidly creative era following the resurrection with the remarkable transformations of the universe which cosmologists tell us happened in the split seconds following the primordial big bang.[13] I have already compared the Christian Church, in its continuing witness to the resurrection, with the background radiation which persists as a 'remembrance' of those early cosmic events. Down the centuries the Church has continued to wrestle with the figure of Jesus, finding expression for its understanding of him phrased in terms of its contemporary experience and world-view. Thus we encounter the imperial figure of the Byzantine Pantocrator, the 'man of sorrows' of the Middle Ages, the 'friend of the poor' of South American liberation theology today. One feels that each of these figures expresses a true insight, but that each fails fully to contain the one to whom they refer. Jesus escapes our attempts

[12] Dunn (1980), p. 253 (a judgement expressed in the original with italic emphasis).
[13] Macquarrie (1990), ch. 3.

to pigeonhole him. Because the Church believes it knows Christ as its living and contemporary Lord, its account of him will always be a dialogue between past tradition and present experience.[14] Pannenberg attributes to Schleiermacher the first self-conscious attempt to construct Christology by way of inference from contemporary Christian experience,[15] but this has always been present as a partial and tacit component in theological thought. Without it, the endeavour would become a merely antiquarian project. Yet contemporary exploration must show respect for historical constraints. It is Jesus, the one who was crucified under Pontius Pilate and who was raised on the third day, with whom we have to deal. Unless we keep this in mind, we shall be liable simply to create Christs in our own image. One of the remarks most frequently quoted in books about Christology is the witty observation that George Tyrrell made about the great German church-historian, Adolf Harnack, that 'The Christ that Harnack sees, looking back through nineteen centuries of Catholic darkness, is only the reflection of a Liberal Protestant face, seen at the bottom of a deep well.' It is a warning of which we need always to be heedful.

So far I have been dealing with what a scientist would call the data, the raw 'facts', of Christology. I put facts in quotation marks because, just as in science we recognize that facts are theory-laden,[16] so here our data are inescapably interpreted, right from the start. They cannot but take confessional form. I have espoused the view that the thinking of the early Church was what Moule calls 'developmental', that 'all the various estimates of Jesus reflected in the New Testament [are], in essence, only attempts to describe what was already there from the beginning. They are not successive additions of something new, but only the drawing out and articulating of what is there.'[17] What was there from the beginning, and continues as part of the contemporary testimony of the Church, is well expressed by John Knox: 'the man Jesus most surely remembered and the heavenly Lord most surely known – and the age-old problem of Christology is implicit in that fact'.[18] In a nutshell, Christians have always been driven to speak of Jesus in terms which call for both the human and the divine. Let me emphasize that the problems of Christology arise not from speculation, but from experience. In Knox's words again, 'the terms "humanity" and "divinity" applied to Christ answer basically, not to ideas and thoughts *about* him, but to the

[14] See Niebuhr (1951) for an analysis of the different forms of interaction between Christ and culture.
[15] Pannenberg (1968), p. 25.
[16] See, for example, Polkinghorne (1986), ch. 2.
[17] Moule (1977), pp. 2–3.
[18] Knox (1967), p. viii.

Church's experience *of* him. There is a divine and a human ingredient in the concrete reality of Christ.'[19]

I think that the titles assigned to Jesus, which we have largely been considering so far, play the role that models do in scientific investigations.[20] They give useful but limited insight. Because their role is frankly heuristic and exploratory, they can be used with a considerable degree of tolerance of unresolved difficulties. The assertion that Jesus is Lord functions in this way, speaking of the authority and transforming power found in the risen Christ, but leaving unsettled the problem of how this is related to the fundamental Lordship of God himself.

Models are useful manners of speaking, but a scientist will not rest content with them alone. What is being sought is understanding, as profound and comprehensive as can be attained. That means looking for a theory. The same quest is inevitable in Christology, though the much greater difficulty of the task (we are concerned, not with the physical world which we transcend, but with the transcendent God and his relation with his creatures) may well make us very modest about the degree of success we are likely to be able to achieve.

Christological theory-making seems to have begun very early, and the first attempt appears to have been what we would now call adoptionism. It is expressed in Acts' account of Peter's sermon at Pentecost: 'God has *made* him both Lord and Christ' (2.36); the resurrection is the divine exaltation of the man Jesus to the status of the heavenly Lord. We find a similar thought expressed in Romans 1.3–4, where Jesus is spoken of as 'descended from David according to the flesh and designated [*horisthentos*, perhaps 'appointed'] Son of God in power according to the Spirit of holiness by his resurrection from the dead'. This way of speaking is not characteristic of Paul, and commentators believe he is quoting an earlier formula, perhaps without due attention to what it might imply (cf. Rom. 8.3, which is more typically Pauline). Adoptionism has at first sight a certain attraction. It neatly divides the human Jesus from the exalted Lord, and its emphasis on the resurrection as God's critical act corresponds to the experience which in the lives of the disciples turned them from puzzled followers into confident proclaimers. It is not without some precedent in the thought of the ancient world, which was familiar with the notion of the post-mortem deification of heroes and emperors, though the ignominious death of Jesus and its universal salvific significance, so that it avails not just for him but for all of us, are features of the Christian story which are without contemporary parallel. It has, however, a fatal theological flaw, for adoptionism assigns the

[19] ibid., p. 54.
[20] See Polkinghorne (1991), ch. 2.

initiative to God only in the last act of the drama, and the Church soon realized that it had to assert that all was from God from the beginning. The exaltation of Jesus could not be pictured as a piece of divine opportunism, trading on the fortunate occurrence of a man worthy of resurrection. God must have been at work in Jesus throughout.

Transference of the moment of adoption from the resurrection to the baptism, or even to the nativity, could not really remedy the matter. In consequence the next development in Christological thought took place, which was to embrace the concept of pre-existence as an expression of God's pre-ordained purpose, so that the earthly life of Jesus was seen as the appearance in history of One who had been part of the divine life and plan from the foundation of the world. In the words of Knox, 'A prologue had been added to the story but the story itself was unchanged.'[21] Pre-existence considerably complicated Christological thinking but it seemed a necessary complication to do justice to the divine initiative. A concept was at hand from Jewish thought which seemed hospitable to this development. It involved the figure of Wisdom (Prov. 8; Ecclus. 24; Wisd. 7), closely associated with God, and indeed his consort in creation, but also closely related to men and coming into encounter with them. Dunn claims that 'The earliest christology to embrace the idea of pre-existence in the NT is Wisdom christology.'[22] We see its full flowering in those three celebrated first chapters of John, Colossians and Hebrews. Yet pre-existence seems also implicit in the Pauline language of God's 'sending' his Son (see also 2 Cor. 8.9), and many of us feel it is present in the pre-Pauline hymn of Philippians 2.6–11 (though others would attribute the language of 'the form of God' and 'equality with God [not] a thing to be grasped' to reference to an Adamic Christology, because they detect echoes of Gen. 3). The scheme of Philippians 2 corresponds to a sequential pattern which Knox calls kenoticism: heavenly figure → man → heavenly figure, which is indeed a prologue added to his characterization of the adoptionist pattern: man → heavenly figure. A certain simplicity is retained, in that, to use later language, one deals with one 'nature' at a time. The difficulty, of course, is the sharpness of the transitions involved. Could there be enough continuity to make it sensible to say that one was talking about the same person?

The addition of a pre-existent stage seems to intensify this problem. The possibility for development of human life is such that it might be felt to be open to post-mortem deification more readily than one can conceive of a heavenly figure being open to becoming a genuine man. We would express the difficulty by saying that the natural precursors of a human being are the

[21] Knox (1967), p. 12.
[22] Dunn (1980), p. 209.

131

genetic and social inheritance into which he or she is born, and how can one square such an origin with the notion of a prior heavenly existence? In its own way, the ancient world felt also a tension between the earthly and the heavenly. One solution that was proposed very early on was docetism: that the true story of Jesus is the story of a heavenly figure who only *appeared* to live a life in human form. This is the one Christological option which is resoundingly rejected by the New Testament writers. People have sometimes alleged that the figure of Jesus in the fourth gospel is so presented as being always totally in control of all circumstances that he is effectively a docetic Christ. I can see that John may have seemed to move dangerously close to that position at times, but I do not think he actually embraced it. One has only to remember that the one who said, 'If any one thirst, let him come to me and drink' (John 7.37), is also the one who says on the cross, 'I thirst' (John 19.28), in order to see that the Johannine Christ is not an invulnerable heavenly figure exempt from the crushing realities of human life. And the Johannine epistles (certainly from the same school of thought, and I believe actually from the same author) are robust in their rejection of docetism: 'every spirit which confesses that Jesus Christ has come in the flesh is of God, and every spirit which does not confess Jesus is not of God' (1 John 4.2–3). The emphatic refutation by the writer to the Hebrews of the idea that Jesus is to be understood in terms of an angelic messenger (Heb. 1.4–2.18) is a similar denial of docetism.

It is in accord with the intellectual temper of our age that the one assertion that we want to make about Jesus without any hesitation is that he was a man. No other thought is possible on our agenda.[23] Docetism holds no temptation for us. (The proper theological necessity for this will be considered later.) Knox is of the opinion that pre-existence is so much at odds with true humanity that, despite the value he places on the kenotic story, he is driven to a kind of eternally purposed adoptionism.[24] I shall later suggest that this desperate remedy is not necessary. Meanwhile, let us consider another assertion traditionally made about Jesus which some have also thought to be inconsistent with his being really human, namely his sinlessness. In the nineteenth century this claim was made the basis of a Christian apologetic which had recoiled from dependence upon miracle or prophecy. The absolute moral purity of Jesus was seen as the ground for his special significance. Today we perceive some difficulties with this idea. Even setting aside the unobservability of inward motivation and desire, it is questioned how such spotlessness can be possible for a man who is truly part of the flawed world of humanity and its social structures. Is there not here

[23] cf. Baillie (1956), ch. 1; Robinson (1972), ch. 2.
[24] Knox (1967), ch. 6.

the danger of a new docetism, given expression (it is alleged) by Paul when he speaks of God 'sending his own Son in the *likeness* [*en homoiōmati*] of sinful flesh' (Rom. 8.3)? Do not such ambiguous incidents as Jesus' relationship with his family (Mark 3.31–5, par.) and with the woman in need who came from outside Israel (Mark 7.24–30, par.) precisely illustrate just such inescapably unsatisfactory entanglements with the limitations of a sinful and imperfect world? I would want to answer that, of course, Jesus lived in the world as it was, with all its inherent contradictions (Heb. 12.3), even to the point of identification, when on his cross he was made, in Paul's deeply mysterious words, 'to be sin who knew no sin' (2 Cor. 5.21), but that all the same he was not tainted by it. That must be an assertion of faith rather than an empirical observation, but it is not an unmotivated assertion for, as John Robinson observes, 'the gospels attribute to Jesus no trace of consciousness of sin or guilt',[25] while the greatest of his followers, from Paul to Francis and beyond, have always been most conscious of the way in which they have fallen short of what they might have been.

So far, the theories about Jesus which we have been considering are of a kind which a physicist would characterize as 'phenomenological'. They are more than mere aggregations of models, but they are too close to what they seek to explain to afford a profoundly satisfying understanding of it. The search for a deeper comprehension of the relation of the human and divine in Jesus occupied the Church in the four centuries up to the Council of Chalcedon, and the celebrated Christological definition of that Council was more an indication of the area for continuing orthodox discussion than the settling of the argument. Knox paraphrases the attitude of the Chalcedonian Fathers in the words, 'we do not see how this could be true, but this is how it is and how it must be'.[26] A quantum physicist can sympathize with an attitude of bafflement at the surprising way things actually turn out to be.

An example of a phenomenological theory in physics would be the wave/particle theory of light, agreeing to use these complementary descriptions in the areas of experimental experience appropriate to their employment. People have often felt drawn to a corresponding complementarity as a way of thinking about divine and human categories applied to Christ. By itself, that is not explanation;[27] it is simply a slogan way of acknowledging that we are driven by Christian experience to a 'two languages' discourse about him. Such talk can separate divine and human nature in a crudely Nestorian way (the divine discerns the marital status of the woman of Samaria; the human asks her for a drink to quench his thirst (John 4)). Through the genius of

[25] Robinson (1972), p. 97.
[26] Knox (1967), p. 100.
[27] See Polkinghorne (1991), pp. 25–8.

Dirac, physics found the deep reconciliation of wave and particle by the invention of quantum field theory. Christology has not yet found its Dirac, and perhaps it is unlikely to do so this side of the *eschaton*, but it has sought to do better than a Nestorian dualism, or an Apollinarian takeover of the human by the divine or an Ebionite suppression of the divine by the human. One is irresistibly reminded of our metaphysical perplexities about how mind and matter are related to each other in ourselves. Complementarity, both where it is understood (wave/particle) and where it is only conjectured (mind/matter), seems always to involve the radicality of reconciling what is regarded as wholly other from the limited point of view of common sense, rather than the assimilation of the vaguely similar. Gorringe comments concerning Christology that

> Because God is the 'Wholly Other' there is not the same contradiction in claiming that Jesus is both human and divine that there would be if we claimed he was both a man and a sheep. Humans and sheep, as [Herbert] McCabe argues, are both members of the same universe; God is not a member of any universe.[28]

His approach to divine presence in Jesus is simply to say that 'God chose to be present as creature',[29] so that the incarnation is seen as a particularly focused mode of immanence.

William Temple once said that 'if a man says that he understands the relation of Deity to humanity in Christ, he only makes it clear that he does not understand at all what is meant by an Incarnation'.[30] There is bound to be an element of inscrutable mystery, which cannot be dispelled, about our talk of God in Christ. Some speculative boldness and some acknowledgement of the limitations of human reason in speaking of the divine are both required for the task. In relation to Christology, Robinson said, 'What myth is to the imagination, metaphysics is to the intellect';[31] we are attempting to sail on seas too deep for simple knowledge.

The Church came to the conclusion that it had to use both divine and human language about Jesus, and it had to do so simultaneously, not sequentially as the primitive Christologies had attempted to do. It was necessary to assert that Christ was true man but also 'of one Being (*homoousios*) with the Father'. The Son of God was eternally divine (not a semi-divine figure, as Arianism suggested, with a beginning before the creation of the world), and when sent by the Father he assumed true

[28] Gorringe (1991), p. 86.
[29] Gorringe (1990), pp. 121–2.
[30] Temple (1924), p. 139.
[31] Robinson (1972), p. 21.

humanity, never more to set it aside. Yet Christ is never equated with God, pure and simple: 'In no serious theology, ancient or modern, has the pre-existent Christ been identified with God, purely and absolutely.'[32] Nor did the manhood of Jesus exist before his conception. A carefully nuanced language is required.

Why are we driven to such incarnational language at all? It is the instinct of a bottom-up thinker to start with a description of the phenomena to be explained. They determine the nature of the problem whose solution is being sought, and therefore they control the nature of the solution which can be regarded as acceptable. The central Christian experience of encounter with Christ is that in him is found a redeeming and transforming power so great that it can only be described in terms of metaphors of a new creation (2 Cor. 5.17) or a new birth (John 3.3–8) or life from the dead (Rom. 6.5–11). No account of Jesus will be convincing to the Christian that is not adequate to explain how this can be. I believe that the work of Christ is the epistemological key to the nature of Christ.[33] Philip Melanchthon said that 'To know Christ is to know his benefits . . .', and that is surely to start in the right place. He continued, '. . . not to contemplate his natures and the modes of his incarnation', but here, I think, he failed to recognize that the one leads naturally to the other. For a critical realist, for whom experience is a guide to what is the case, the salvific encounter with Christ must open up the question of who it really is that we have encountered.

God alone can redeem us from the entail of sin and the flawed life which results from our alienation from him. Only divine action can save us, and 'the agent must really be there in the event'.[34] Vernon White proposes a 'criterion of moral authenticity': 'unless and until God himself has experienced suffering, death, and the temptation to sin, and overcome them, as a human individual, he has no moral authority to overcome them in and with the rest of humanity'.[35] Human redemption comes about through divine involvement, and not by an act of divine magic. The incarnation is the narrow point on which the large claim for universal salvific validity stemming from a particular life and death must balance. The human condition is such that it cannot be dealt with simply by an authorized representative (the Hebrew idea of *shaliach*), however inspired, but it requires actual divine participation. It is therefore essential, if Jesus is the Saviour, that God is fully present in him throughout. In Athanasius' words, 'He became man that we might become divine,' so that we might share in

---

[32] Knox (1967), p. 109.
[33] Polkinghorne (1988), ch. 6.
[34] White (1991), p. 38.
[35] ibid., p. 39.

135

the life of God, and consequently that life of God must be in him. Yet the Redeemer is not a gnostic Christ imparting the secrets of divine wisdom, who could indeed be a heavenly figure in human disguise. The mystery of our redemption is something altogether deeper than that. It proceeds, not from the outside by illumination, but from the inside by participation. We need transformation, not information. That is why docetism is so totally unacceptable to Christian thought. The Saviour must be truly and fully human. In Gregory of Nazianzus' famous words, 'what is not assumed is not redeemed'. A heavenly figure could be of no redemptive significance for us. We would have no share in him. John Knox poses the dilemma: 'How could Christ have saved us if he was not a human being like ourselves? How could a human being like ourselves have saved us?'[36] Later he rephrases the questions, 'Who could have saved us but God himself? How could even he have saved us except through a human being like ourselves?'[37] Seeking the response to these questions is the driving force of Christological thought. Pannenberg says that 'Since Schleiermacher the close tie between Christology and soteriology has won general acceptance in theology,'[38] but he expresses a certain wariness lest Jesus should become just a symbol for the fulfilment of human hopes and wishes. We must remember that 'Jesus possesses significance "for us" only to the extent this significance is inherent in himself, in his history, and in his person constituted by his history.'[39] I entirely agree. Rudolf Bultmann posed the question of Jesus, 'Does he help me because he is God's Son, or is he the Son of God because he helps me?' Thiselton is surely right to reject what he calls this presentation of 'thoroughly false alternatives'.[40] There is a kind of 'soteriological circle' in which Christ's being and function are interdependently related.

The New Testament writers see a particular saving significance in the death of Jesus: 'God shows his love for us in that while we were yet sinners Christ died for us' (Rom. 5.8); 'He himself bore our sins in his body on the tree' (1 Pet. 2.24). Nowhere is the gap between the testimony of Christian experience and the adequate theological understanding of that experience greater than in relation to the atoning death of Christ. Theories as manifestly inappropriate as the deception of the devil (who tried to capture the sinless one on whom he had no legitimate claim), or the propitiation of an angry God, or the satisfaction of the affronted dignity of the feudal Lord of the universe, have all been canvassed, as have theories as weakly subjective as an exemplary action or as problematically mythological as victory over the

[36] Knox (1967), p. 52.
[37] ibid., p. 92.
[38] Pannenberg (1968), p. 38.
[39] ibid., p. 48.
[40] Thiselton (1980), pp. 267, 269.

Powers. Let me stress that, however unsatisfactory these accounts may be if regarded as theories, they are attempts to use models[41] to understand a phenomenon (redemption through the cross of Christ, the forgiveness of sins and the promise of saving transformation) which has been fundamental to Christian experience throughout the centuries. I agree with Frances Young that 'response to the gospel of redemption is primary, and expositions of this in theological terms are secondary'.[42] I would only want to make two points.

The first is that the New Testament sees the cross as an act of *God*: 'God was in Christ reconciling the world to himself' (2 Cor. 5.19). In Baillie's words, 'God's merciful attitude towards us is never regarded as the *result* of the process, but as its cause and source.'[43] Young summarizes the view of Athanasius expressed in the *De Incarnatione* as being that 'The sacrifice of Christ . . . was a sort of "self-propitiation" offered by God to God to make atonement for the existence of evil in his universe.'[44] The cross is the eruption into history of the love of God shown in the Lamb slain from the foundation of the world (Rev. 13.8). This understanding is a powerful motivation for the search for an incarnational theology as opposed to an adoptionist theology. It is also of profound significance in relation to that most agonizing of all theological problems, the problem of suffering, for if God was in Christ in his death, then he is truly a 'fellow-sufferer', knowing the bitterness of the human condition from the inside. 'In the lonely figure hanging in the darkness and dereliction of Calvary the Christian believes that he sees God opening his arms to embrace the bitterness of the strange world he has made.'[45] Brian Hebblethwaite expresses a similar thought, 'Only if we can say that God has *himself*, on the Cross, "borne our sorrows" can we find him universally present "in" the sufferings of others.'[46] Moltmann says of Jesus, 'He became the kind of man we do not want to be: an outcast, accursed, crucified . . . When the crucified Jesus is called "the image of the invisible God" the meaning is that *this* is God, and God is like *this*.'[47] Such a thought makes a powerful appeal in the century of Auschwitz, presenting to us the crucified God as 'the basis for a real hope which both embraces and overcomes the world, and the ground for a love which is stronger than death and can sustain death'.[48]

[41] Polkinghorne (1991), ch. 2.
[42] Young (1975), p. 6.
[43] Baillie (1956), p. 188.
[44] Young (1975), p. 93.
[45] Polkinghorne (1989a), p. 68.
[46] Hebblethwaite (1987), p. 36.
[47] Moltmann (1974), p. 205.
[48] ibid., p. 278.

The second point I want to make addresses the question of why atonement is linked with a death. It signifies, at the very least, that forgiveness and reconciliation are costly. That is why the Christian tradition has used sacrificial language about the death of Christ. What is involved is not an easy divine dismissal of evil as a not very important matter, readily disposed of. We know in our own experience that true forgiveness – forgiveness that acknowledges the fault and its hurtfulness – is painful and dear. Baillie asks the rhetorical question of God's forgiveness, 'Does the whole process of reconciliation cost Him nothing? Is his forgiveness facile and cheap?'[49] It cannot be, and the cross of Christ exhibits its intrinsic costliness. Our reconciliation to God is both objective (Christ does something for us that we could not do for ourselves) and subjective (we have to embrace and make our own the forgiveness made available in him). A proper balance is maintained between the two if we recognize with Frances Young that the objective language of sacrifice finds its proper usage within the worshipping community of the redeemed: 'Sacrifice is properly treated as cult-language, not the language of law-courts and judgements; in this context, Christ's act is seen as a sacrifice in which Christians have to partake in order to receive its benefits.'[50] The prime contexts for this cultic participation are the dominical sacraments of baptism and the Eucharist.

The early Christian teaching says that 'Christ died for [*huper*, on behalf of] our sins in accordance with the scriptures' (1 Cor. 15.3). Perhaps one of the scriptures those first believers had in mind was Isaiah 53, where God's Servant shall 'make many to be accounted righteous; and he shall bear their iniquities' (v. 11). How can one man avail for the many in this way? It is certainly a fundamental claim of the New Testament that Jesus died 'for us' (Rom. 5.8, etc.). Grenz explains an important distinction between two different ways in which that death might be on our behalf: 'In exclusive substitution the suffering a person experiences in the place of another means that the other need not suffer the same fate. Inclusive substitution, in contrast, means that the substitute shares the situation of the others (death in the case of Jesus) and thereby alters that situation.'[51] It is the latter idea of a participatory transformation – not just of death but of alienation from God (Mark 15.34; 2 Cor. 5.21; Gal. 3.13) – which I find most helpful. Jesus did not pay a debt, but he redeemed an experience by sharing it.

But how could the participation of the one avail for the many? I do not think that we can make full sense of the Christian understanding of Christ's death without being prepared to embrace the notion of a degree of human

---

[49] Baillie (1956), p. 172.
[50] Young (1975), p. 96.
[51] Grenz (1990), pp. 121–2.

solidarity, a notion so foreign to our individualistic, atomized, Western thinking, though it would not be so strange, I suspect, to someone from tribal Africa.[52] There must be a human aspect, as well as a divine aspect, to the way in which the particular event of the cross has universal significance. The New Testament sets before us the idea of the solidarity of the redeemed 'in Christ' (en Christō(i)), that oft-repeated Pauline phrase (Rom. 6.11; 8.1; etc.) and it expresses it in many ways: the Church as the body of Christ,[53] Christ as the true Vine (John 15.1–11), the corporate overtones of the phrase 'the Son of man'. The concept is difficult, and certainly demands for its adequate understanding more than the acknowledgement that there is an inescapable social dimension to every human life. Jesus is the ground of the Church's existence, not just in the first generation but in every generation. While Knox says, 'The Church is the *distinctive* Christian reality,'[54] he also acknowledges that its existence

> came into being around *him* and because of *him*; he was its creative centre and largely determined its concrete character. Moreover the memory of him as an actual human being and the continuing awareness of him as a Victorious Spirit are still, and always will be, centrally constitutive of the Church.[55]

I would wish to carry the implications of this further and follow Moule who, in his careful account of corporate language about Christ, says that in Paul's writings he 'was found to be an "inclusive" personality. And this means, in effect, that Paul was led to conceive of Christ as any theist conceives of God: personal, indeed, but transcending the individual category.'[56] Here, again, the human and the divine meet in Christ.

The saving work of Christ and his corporate character are attested by Christian experience and, in my view, enforce necessary conditions which any adequate Christology must succeed in meeting. I think that evolutionary Christologies, which speak of Jesus as the 'new emergent',[57] signally fail to pass this test. The idea seems to be at least as old as Schleiermacher.[58] It also owes something to Hegel and something to Darwin, and it has been particularly popular among modern writers from the science-and-religion school. Peacocke wishes to speak of 'incarnation' in terms of the possibility that 'a man [may] so reflect God, be so wholly open to God, that God's

[52] cf. Donovan (1982).
[53] Robinson (1952).
[54] Knox (1967), p. 66.
[55] ibid., p. 89.
[56] Moule (1977), p. 95.
[57] Theissen (1984).
[58] Macquarrie (1990), pp. 206–8.

presence was unveiled to other men in a new, emergent, and unexpected manner',[59] and he later speaks of Jesus as 'emergence-from-continuity' so that he is 'a unique manifestation of a possibility always inherently there for man by his potential nature'.[60] Peacocke embraces a kind of Pelagian Christology. Barbour suggests that 'we may view both the human and the divine activity in Christ as a continuation and intensification of what had been occurring previously. We can think of him as representing a new stage in evolution and a new stage in God's activity.'[61] One might ask whether the next stage in evolution will regularly produce beings who rise from the dead? Torrance says that 'Western thought from ancient times in Greece has developed the habit of getting rid of singularities through some kind of scientific horror of unique events.'[62] We must overcome that horror if we are to do justice to the Christ-event. Jesus resists assimilation to general categories. Something much more mysterious and hopeful happened in him than can be contained in evolutionary terms. How could a 'new emergent' be of use to us, who have manifestly failed to emerge? The insight of continuity with humankind, which these evolutionary Christologies are striving to express, is true, but its location is misplaced. Our hope lies not in an encouragement to make more of the potentialities of present process, but in a call to participate in Christ in that eschatological transformation constituting the new creation, which is to grow from the seminal event of his resurrection.

Similar difficulties afflict Christologies which place the emphasis on Jesus as a 'window into God'. John Robinson wished to assert that

one who was totally and utterly a man – and had never been anything other than a man or more than a man – so completely embodied what was from the beginning the meaning and purpose of God's self-expression . . . that it could be said, and had to be said, of that man, 'He was God's man', or 'God was in Christ', or even that he was 'God for us'.[63]

No account of the incarnation will fail to emphasize that Jesus makes God known to us in the plainest possible terms, by living the life of a man, but to rely on the revelatory character of his life alone is to adopt a gnostic account

[59] Peacocke (1979), p. 213.
[60] ibid., pp. 241–2.
[61] Barbour (1990), p. 211.
[62] Torrance (1980), p. 100.
[63] Robinson (1972), p. 179. Cullmann (1963) characterizes the New Testament attitude to Christ as being that he is God in his self-revelation, but this judgement arises from Cullmann's concentration on titles. The New Testament speaks much of salvation, but only infrequently uses the title 'Saviour'.

of our redemption. Let me say again that I believe our need is for transformation, not just information. Jesus is the second Adam (the seed of a new humanity), not the second Moses (the conveyer of a new revelation).[64]

No more successful are inspirational Christologies – that see Jesus as being a man who was open to divine inspiration to an unprecedented degree – and for the same reason, that they do not explain how that could avail for us. Such functionalist Christologies, framed in terms of enhanced divine activity through Jesus, fail to explain how we are to follow 'the perfect pupil'.[65] They also fail to do justice to the deep Christian intuition of the uniqueness of Christ. Brian Hebblethwaite says, 'The weakness of exemplar-ist theories of atonement is not just their subjective nature. It is much more their inability to account for the absoluteness of the example.'[66] He goes on to criticize Lampe's inspirational Christology[67] on precisely this point. Finally, such Christologies lose the profound understanding of God's involvement with the human predicament that incarnational Christologies preserve. Gorringe says that 'The issue between Lampe and those who hold to a more incarnational christology is whether we live by the story of human *possibility*, enabled by grace, or whether we live by the story of divine involvement and suffering.'[68]

Thus I am driven by the necessity of finding a Christology adequate to the Christian experience of new life in Christ, to seek an understanding along the orthodox and traditional lines which speak in terms of the presence of both true humanity and true deity in Jesus. Yet how can God be present in the life of a man, however remarkable? Some focusing of the Infinite upon the finite, a projection from the spacious realm of deity into the confined regime of humanity, must be involved. That language is suggestive of a kenotic Christology, the laying aside of divine attributes in an accom-modation to the realities of human limitation.[69] One of the difficulties with such an idea has always been its consistency with continued divine governance of the universe. In William Temple's direct if somewhat naive words, 'What was happening to the rest of the universe during the period of our Lord's earthly life?'[70] Baillie produces somewhat more sophisticated difficulties when he asks questions going beyond the earthly life of Jesus: How permanent was the assumption of humanity? Was there just a

[64] Herein lies the essential distinction between an ontological incarnational Christology and one that is merely functional.
[65] Brown (1985), p. 237.
[66] Hebblethwaite (1987), p. 63.
[67] Lampe (1977); cf. Ward (1991).
[68] Gorringe (1990), pp. 46–7.
[69] See Macquarrie (1990), pp. 245–50; Baillie (1956), pp. 94–8.
[70] Quoted in Baillie (1956), p. 96.

temporary theophany? Were the divine attributes eventually recovered?[71] These problems all arise from an assumption that the *whole* of the Second Person of the Trinity was involved in the kenotic episode of the incarnation. It seems to be suggested that the divine Person was, as it were, totally in abeyance during the episode of the incarnation, corresponding to the primitive kenotic scheme: divine → human → divine, referred to earlier (p. 131). That seems to be an unnecessarily crude idea. There have been other views put forward, such as the so-called *extra calvinisticum*. John Calvin taught that during the earthly life of Jesus, the *Logos* 'continuously filled the world even as he had done from the beginning'.[72] I think such thoughts are worth pursuing, and that a dipolar account of God's relation to time and eternity (chapter 3) helps one to conceive of some helpful new possibilities. I have had the temerity to suggest that it is the temporal pole of the Son that is involved in the incarnate life of Jesus.[73] That is as far as a bottom-up thinker dares to go in speculating on such profound theological matters. It is, I believe, a proposal fully consistent with the soteriological considerations which have dominated this chapter, because for us embodied (and so necessarily temporal) beings, our future sharing in the life of God could only be within the unending exploration, in the time of the world-to-come, of the inexhaustible riches of the temporal pole of the divine nature. All the divine that could be shared by humanity is united with humanity in Christ. He is *totus deus* (wholly God), but the earthly Jesus is not *totum dei* (all of God).

The thrust of this argument-from-below has been to start with the human figure of Jesus and to recognize that adequate talk of him is driven to employ also the language of the divine. Thomas Morris has presented a discussion-from-above in which he argues that 'the Christian claim is that in order to be fully human it is not necessary to be merely human',[74] so that he adopts a 'two minds view of Christ',[75] in which the infinite omniscient divine mind 'encloses' a finite partly-ignorant human mind. Against kenoticism, he argues that for Christ 'It is not by virtue of what he gave up, but in virtue of what he took on, that he humbled himself.'[76] Only the merely human has to come to be and is contingent. In Morris's view, the fully human is compatible with necessary being, true incarnation with divine infinity, in a straightforward way. My difficulty here would be the compatibility of this with soteriological considerations; if human contingency is not completely

---

[71] ibid., 94–8.
[72] Quoted in Macquarrie (1990), p. 171.
[73] Polkinghorne (1989a), ch. 8.
[74] Morris (1986), p. 65.
[75] ibid., p. 102.
[76] ibid., p. 104.

assumed, how is it redeemed? It does not seem sufficiently embraced if historicity is just added in parallel to the divine. A greater immersion of God the Son in the human seems called for. The notion of the kenotic self-abasement of the divine temporal pole, alluded to earlier, is a groping attempt at a more integrated account.

Morris's strategy is designed to avoid a head-on contradiction between divine necessity and human contingency by setting them side by side in a way which, despite the author's disclaimer,[77] still seems Nestorian. His approach is powerfully and logically set out, but I instinctively feel that Christology demands a more radical revision of our modes of thought, possibly involving polarity of divine necessity/contingency[78] or even (whatever that might prove to mean) divine/human complementarity in Christ. The puzzling paradox of the wave and particle nature of light was not resolved by clarifying classical thinking, but by transforming it into a new quantum mode of thought.

I would like to conclude with something I wrote previously about the presence of divine Sonship in Jesus:

> The power of myth and the power of actuality fuse in the incarnation. What could be more profound than that God should take human form, make himself known in human terms, share the suffering of the strange world he has made and on the cross open his arms to embrace its bitterness? That is a story which moves me at the deepest possible level. Yet it is no tale projected on to a shadowy figure of ancient legend. It is concerned with what actually happened in the concrete person of Jesus Christ, a wandering teacher in a particular province of the Roman Empire, at a particular point in history. The centre of Christianity lies in the *realized myth* of the incarnation.[79]

'he became incarnate of the Virgin Mary and was made man'

The presence of a true divine initiative in the life of Jesus, and at the same time his true continuity with humanity, have together found traditional expression in the doctrine of the virginal conception, in which the overshadowing of the Spirit and the obedience of Mary play their appropriate parts in the origin of the human Jesus. Modern thought has tended to find here more tension than resolution. If such mixed 'parentage'

[77] ibid., pp. 154–8.
[78] Ward (1982a). The motivation for Ward's discussion comes principally from a consideration of divine creation.
[79] Polkinghorne (1988), p. 97.

is taken seriously as a miracle of God – and it would be quite as radically 'unnatural' an event in terms of everyday experience and expectation as is the resurrection – does it not divorce Jesus from the rest of humanity? The problem is compounded by a consideration of the scriptural evidence. The resurrection of Jesus is referred to in every major New Testament writing, but the virginal conception is much less widely attested. Notoriously, Paul makes no apparent reference to it (cf. Gal. 4.4: 'born of woman, born under the law'). Explicitly, the story is given only in the two familiar narratives of Matthew (1.18–25) and Luke (1.26–38), of which the former has the plainer speaking about the virginal character of Jesus' conception. The nativity narratives are very different in the two gospels (magi and a flight into Egypt in Matthew, shepherds in Luke), and they have features suggestive of legendary accretion (moving stars and angelic hosts). Yet these very differences indicate two variants of tradition which nevertheless have a common core concerning the remarkable character of Jesus' conception and birth.

It is scarcely surprising, in view of such problems, that even careful and conservative scholars exhibit a reserve about the virginal conception. Concerning the circumstances of the nativity, Dodd writes, 'That there is a basis of fact somewhere behind it all need not be doubted, but he would be a bold man who would presume to draw a firm line between fact and symbol'.[80] On the virginal conception, Moule says, 'I am not prepared to take sides in that debate.'[81] No doubt, the confession of the creed that Jesus was born of the Virgin Mary is the assertion which gives the greatest trouble to the bottom-up thinker.

Yet there is something more to be said. On the evidential side there are some modest signs elsewhere in the gospels that there was something peculiar about Jesus' birth. In Mark 6.3 he is described as 'son of Mary', which is totally contrary to Jewish usage, which habitually associated a son with his father, though there is a variant reading of the text which refers to 'son of the carpenter and Mary' (itself an odd expression). In John (where so much is alluded to rather than explicitly spelled out), the Jews say pointedly to Jesus, 'We were not born of fornication' (8.41). In the Matthean genealogy (Matt. 1.2–16) there is the egregious appearance of 'doubtful' women such as Tamar and Rahab in this predominantly male lineage. Robinson concluded, on the basis of these indications, that 'The only choice open to us is between a virgin birth and an illegitimate birth.'[82] A certain type of modern mind can find it quite appropriate to suppose that Jesus was

[80] Dodd (1961), p. 42.
[81] Moule (1977), p. 5.
[82] Robinson (1984), p. 4.

born as the fruit of an illicit union, but I cannot share the view that this is so.[83]

Such considerations, together with the symbolic appropriateness of the fusion of divine initiative and human co-operation which the virginal conception would signify as the origin of Emmanuel ('God with us'), along with the recognition, already referred to, that the incarnation is the union of the mythical and the historical, persuade me that the words 'born of the Virgin Mary' can be a proper part of the creed of a bottom-up thinker.

[83] cf. the essay by H. A. Williams in Vidler (1962).

8

The Spirit and the Church

'The Holy Spirit, the Lord, the giver of life, . . .
with the Father and the Son he is
worshipped and glorified'

The early Church felt that it experienced divine power present within it
with a peculiar intensity and personality. There was language available with
which to describe this experience, the language of spirit (*ruach*), used in the
Old Testament in relation to creation (Gen. 1.2), the empowerment of
individuals (Exod. 35.30–36.1), the continuance of life (Ps. 104.29–30), and
the fulfilment of the age to come (Ezek. 36.26–7; Joel 2.28–9). In both
Greek and Hebrew the word for spirit means also 'breath' or 'wind', and the
expansiveness of its allusion fitted the word to cover many applications.
Common to them all is certainty of effect together with invisibility of
presence.

Acts describes the day of Pentecost as being the start of the indwelling of
God's Spirit in some new way, more continuous and more manifest than had
been experienced before (Acts 2.1–36). The resurrection appearances had
come to a close but, to use the Johannine language for this phenomenon, 'the
Counsellor, the Holy Spirit . . . the Spirit of truth' had now been sent to
continue Christ's work (John 14.26; 16.12–15). Other special givings of the
Spirit are mentioned (Acts 8.14–17; 10.44; 19.6), but his presence and
activity are implicit throughout the whole book of Acts and, indeed, the
New Testament generally.

As later generations became more systematically reflective upon Christian experience, some perplexity and controversy arose about what was a
fitting theological characterization of this phenomenon. How specifically
divine and how specifically personal did the language need to be? Was the
experience one of a utilizable gift of power or of an enabling divine presence?
It took several centuries to reach the agreed conclusion that the Holy Spirit
is to be spoken of as a third divine Person, together with the Father and the
Son. The version of the Nicene Creed adopted at the Council of
Constantinople in 381 enshrines this understanding. Even so, a residual

reserve seems to have continued for, though the Spirit is given the title of 'Lord' and accorded the divine honours of being 'worshipped and glorified', yet he is not explicitly stated to be 'one in Being with the Father', as is asserted of the Son. In much Christian practice this reserve has been perpetuated. Many believers would confess that their notions of the Holy Spirit are decidedly hazy, and Pentecost is that major festival in the Church's year which attracts the least attention. In Christian art the symbol of the Spirit as a dove contrasts with the strongly personal figures of the kingly Father and the crucified Son, and in some medieval representations of the Trinity even this degree of explicit presence of the Spirit is missing.

I believe that this elusive character of our thought and expression about the Spirit is the reflection of a deep theological truth. He is the *deus absconditus*, the hidden God, because his working is from within; his is the divine presence ever-active in the unfolding process of the created world and never wholly disentangleable from that process. In John V. Taylor's words, he is 'God working anonymously and on the inside: the beyond in the midst.'[1] In the world of our personal encounter he is the third party whose unseen presence is the enhancer of our meeting. 'We can never be directly aware of the Spirit, since in any experience of meeting and recognition he is always the go-between who creates awareness.'[2] In the economy of our spiritual life, the Spirit is the one in whom we pray (Rom. 8.26), through the Son to the Father. Moltmann says of those who pray thus, 'Their cries are uttered in the presence of God and in their own situation.'[3] The Spirit is God with us in the particularities of our lives and with the necessary discretion of a love which does not overwhelm those on whom it is bestowed. 'God's absence in his presence is not merely an estrangement. It is liberation too.'[4]

This self-effacing role of the Spirit finds luminous expression in the thought of the Eastern Church. Vladimir Lossky writes:

> The divine Persons do not themselves assert themselves, but one bears witness to another. It is for this reason that St John Damascene said that 'the Son is the image of the Father, and the Spirit the image of the Son'. It follows that the third Hypostasis of the Trinity is the only one not having His image in another Person. The Holy Spirit, as Person, remains unmanifested, hidden, concealing Himself in His very appearing.[5]

[1] Taylor (1972), p. 5.
[2] ibid., p. 43.
[3] Moltmann (1992), p. 288.
[4] ibid., p. 205.
[5] Lossky (1957), p. 160.

Lossky goes on to quote St Symeon the New Theologian, 'come, hidden mystery . . . thy name so greatly desired, and constantly proclaimed, none is able to say what it is'. Yet Gregory Nazianzus said: 'The Old Testament manifested the Father plainly, the Son obscurely. The New Testament revealed the Son and hinted at the divinity of the Holy Spirit. Today the Spirit dwells with us and makes Himself more clearly known.'[6] There is here a notion of progressive revelation (cf. John 16.14–15) which accords with the Church's dawning recognition in the first centuries of the true theological status of the Spirit. Though he is the hidden one, on the inside, that cannot amount to complete ineffability, otherwise there would literally be nothing to say.

One of the earliest perplexities was how the Spirit was related to Christ and God. Is one dealing with (or, better, being dealt with by) the Spirit of God in a straightforward extension of the Old Testament sense, or the Spirit of Christ in the sense of his continuing diffused activity following the ascension, or are we involved in encounter with a third divine Person? Not surprisingly, there was a good deal of initial confusion. In one verse in Romans (8.9) Paul appears to use *pneuma* in all three senses. Eventually, of course, it was decided that it was in terms of the third divine Person, closely associated with the Father and the Son but not to be identified with them, that the truest account of Christian experience could be expressed. One reason for accepting this lies, I think, in the different kinds of encounter with God which seem to be involved. The meeting with Christ is in sharp focus; the meeting with the Spirit is in softer focus. 'Spirit is experienced as inspiration, Word as revelation,'[7] is Taylor's way of putting it.

Lossky draws our attention to another distinction between Word and Spirit: 'Christ becomes the sole image appropriate to the common nature of humanity. The Holy Spirit grants to each person created in the image of God the possibility of fulfilling the likeness of the common nature.'[8] The one is the inclusive destiny of us all; the other is the enabler of our individuality within that destiny. When the tongues of fire descended at Pentecost they are represented as being 'distributed [*diamerizomenai*, divided] and resting on each one of them' (Acts 2.3). When Paul speaks of the gifts of the Spirit, he speaks of their variety among those to whom they are given (1 Cor. 12.4–11). Here is an important clue to why we speak of the Spirit as a person, and not just an impersonal force or power. He deals individually with different people. Taylor says, 'That which was encountered was not simply energy but also form; not simply drive, but also direction; not only holy, but also

[6] Quoted in ibid., p. 161.
[7] Taylor (1972), p. 61.
[8] Lossky (1957), pp. 166–7.

righteous.'[9] I would add, 'not simply general but also particular'. The One who is the creator of the community of the Church must be personal, that is able to meet people both in their individuality and in their relatedness.

Another reason for declining the language of force or power for our speaking about the Spirit is the insight that he participates not only in action but also in passion. Paul speaks of the creation groaning and ourselves groaning with it and then goes on to speak of the Spirit interceding with 'sighs [better, 'groans', for it is the same Greek root] too deep for words' (Rom. 8.22–3, 26; cf. Eph. 4.30).

It could not be otherwise in the costliness of true personal encounter. Taylor sees the Spirit's hidden presence as the ground of those experiences of disclosure in which we seem to penetrate beyond the surface into the heart of things, moments able 'to turn an *It* into a *Thou*'.[10] The Spirit is, therefore, the unseen enabler of the truly personal, the hyphen in the I-Thou relationship.[11] He who does not have his divine image within the Trinity, is at work creating that image through the transformation of humanity. In Lossky's words about eschatological destiny: 'It is then that this divine Person, now unknown, not having His image in another Hypostasis, will manifest himself in deified persons: for the multitude of the saints will be his image.'[12] That the Spirit is the agent of sanctification is one reason for the necessity of asserting his divinity, for only the fully divine can fully save and transform.[13]

Another consequence of the Spirit's divinity is that, however particularly and with due recognition he may be seen at work within the Church, his action cannot be confined to its history alone. God is not the sole possession of any group. If, 'When the Spirit of truth comes, he will guide you into all the truth' (John 16.13), then we must surely understand this as implying his activity in the questing of all truth-seeking communities, including the spiritual explorations of other religious traditions and the physical explorations of the scientific tradition.

The glimpsing of the 'beyond-in-the-midst' within the order of creation has found recurring expression in Christian understanding. One thinks of Celtic and Franciscan spiritualities, with their closeness to the natural world. At first sight, such insights of the Cosmic Spirit might seem to have become problematic in the light of our modern knowledge of physical process. If that were so, it would be a severe difficulty, for it would not be theologically

[9] Taylor (1972), p. 60.
[10] ibid., p. 17.
[11] ibid., p. 29.
[12] Lossky (1957), p. 173.
[13] cf. Pannenberg (1968), pp. 172–4.

satisfactory to confine our expectation of the Spirit's work to its lying within the depths of the human psyche alone. Pannenberg puts the issue squarely before us:

> Contemporary theology lacks a doctrine of the Holy Spirit that corresponds in breadth to the Biblical concept of the Spirit. Such a doctrine would require a treatment of our present knowledge of the causes of life. Can we still speak today of a 'spiritual' origin of all life? What sense would such talk have with respect to the phenomena and structures of life that have been explored by biology?[14]

Taylor's response would be that

> if we think of a Creator at all, we are to find him always on the inside of creation. And if God is really on the inside, we must find him in the processes, not in the gaps. We know now there are no gaps, no points at which a special intervention is conceivable. From first to last the process has been continuous. Nature is all of a piece, a seamless robe . . . If the hand of God is to be recognized in this continuous creation, it must be found not in isolated intrusions, not in any gaps, but in the very process itself.[15]

I think the insights of modern science in fact help us to understand what that might mean.[16] To reject the idea of a fitful divine interventionism is not to embrace the deistic notion of a self-contained universe left to the unfolding of its inevitable history. The concept of divine interaction within cosmic process, through the input of information into its flexible and open development, which is the view put forward in these lectures, is perfectly consonant with the activity of the Spirit 'really on the inside'.

However hidden the work of the Spirit may be, and however difficult the framing of an adequate theological account of his divine nature, I think the effort to articulate these matters is one which must be made. Too often the idea of the Spirit has been used as an unfocused expression of a general divine presence, without sufficient consideration of what that might mean. Writers of the science-and-religion school have not been exempt from this shortcoming. The Spirit must be more than a religious cipher for the scientifically discerned 'optimistic arrow of time', leading from simplicity to complexity. He must be truly at work within that unfolding fruitfulness, drawing the universe on to purposed levels of fulfilment. Therefore, it does

[14] ibid., p. 171.
[15] Taylor (1972), p. 28.
[16] Polkinghorne (1989a); see also, ch. 4 above.

not seem quite enough to say with Arthur Peacocke, 'The continuing creative power which is manifest as a nisus at all levels of existence to attain its intended form is, in the Christian tradition, God as "Holy Spirit".'[17] The precarious fertility of cosmic history is not just the outcome of a drive towards complexity, but on its inside is the passion and action of the personal Spirit. The hiddenness of his working makes that statement an assertion of faith, but one that is motivated by the recognition that the coming-to-be of created personality is the most significant event of cosmic history.

Distinctions must not be blurred, nor the Spirit used as merely a convenient religious common denominator. I have already protested against an inspirational Christology, and I am not happy, as Peacocke seems to be, with a situation in which 'No attempt is then made to distinguish between the being of God revealed in Jesus and that which might be manifest in us too.'[18] (Similar thoughts are to be found in Barbour.[19]) If the Spirit is the *beyond*-in-the-midst, his presence and activity must be more than general experience regarded through theologically tinted spectacles. Peacocke (at least in his earlier writings) and Barbour are both inclined to a panentheistic account, which partially assimilates the divine and the cosmic to each other. I believe, on the contrary, that it is important to maintain a clear distinction between the Creator and his creation[20] (not least, as I shall discuss in the next chapter, if God is to be the ground of a hope independent of present physical process). The work of the Spirit in continuing creation may not be totally disentangleable from unfolding physical process, but I do not think it is simply to be identified with it. The work of the Spirit in our lives may not be totally disentangleable from our personal decisions, but it is the paradox of grace that the two are again not simply to be identified.

In traditional Christian theology, only the Son is incarnate, taking the form of a creature by becoming truly man. The Spirit is not incarnated in the cosmos. Pneumatology and Christology must not be assimilated to each other. To blur the distinction between Creator and creation in a panentheistic way is to fail to do justice to the experience of God's irreducible otherness, to intensify the problem of evil and to jeopardize the concept of Christian hope, replacing it by an illusory evolutionary optimism. I find it easier to speak negatively than positively; to articulate what I cannot accept more clearly than what I want to assert. The Spirit's work is concealed within the flow of present process, but its power derives from the

[17] Peacocke (1986), p. 125.
[18] Peacocke (1979), p. 242.
[19] Barbour (1990), pp. 176–8 and 209–14.
[20] Polkinghorne (1989a), ch. 2.

presence of God's future within that process. Were the Spirit not also transcendent, his immanence might be no more than Spinoza's identification of God and nature. As it is, the Spirit is the pledge (*arrabōn*) of future fulfilment, 'given us . . . as a guarantee' (2 Cor. 5.5). I shall return to this point later.

'he has spoken through the prophets'

The role of the Spirit as the inspirer of scripture was considered by the Fathers to be sufficiently important to merit its own sentence in the creed. I have taken the opportunity elsewhere to consider the nature and use of the Bible,[21] and I do not want to repeat the discussion here. Once again we find an entanglement of the Spirit with creation which will not be susceptible to precise unravelling. Scripture arises from inspiration, not dictation; it mixes continuing truth about the divine with the culturally limited expression and understanding of its authors. The claim that the New Testament was inspired by the Spirit is reinforced by the consideration that the brief period of its writing saw the emergence, in Paul and John and the writer to the Hebrews, of three thinkers whose creative insight has not been surpassed in the subsequent history of theological thought.

The Bible is not a kind of divinely guaranteed textbook in which we can, without any trouble, look up all the answers. I find the notion of the 'classic',[22] rooted in its own age but possessing through its underlying universality the power to speak across the centuries to other ages, to be the category which best contains my own understanding of the spiritual power of scripture. Yet that classic is not just a tale told to illustrate general truths, but it has its anchorage in the particularities of history. The Bible records the foundational events of Israel's experience of God, and the life and death and resurrection of Jesus Christ, which are the indispensable base on which Christian understanding is to be built. The account thus presented is not a mere chronicle of occurrences, but the story is told in a way that makes its inner meaning accessible, revealing what was happening on the inside of what was going on. (The gospels have a biographical basis but they are not biographies.) It is the power of a classic, and pre-eminently the power of that great spiritual classic, the Bible, to speak freshly to each generation of its readers, revealing further truth, which becomes available, not by a process of alien imposition, but by continuing elucidation of the profundity of the text. In our contemporary encounter with the ancient writings, understanding

[21] Polkinghorne (1991), ch. 5.
[22] cf. Tracy (1981).

requires the search for a creative fusion of the 'two horizons'[23] of author and reader. From the tension between past and present can come a deeper insight which neither denies the strangeness of the past in its cultural distance from us nor is cramped by the confines of a purely twentieth-century perspective. Thiselton says of the hermeneutical task that 'We need *both* distancing *and* an *openness* to the text which will yield progress towards the fusion of horizons.'[24] In that way we might hope to achieve what Schleiermacher thought was the paradoxical object of hermeneutics, 'to understand the writer better than he understood himself'.[25]

There are two further things I would want to say. One is that the Spirit's activity in relation to scripture cannot be confined to the initiating moments of authorship. He must be conceived as being at work in the Church's endorsement leading to the formation of the canon, and in the developing understanding of the sacred writings within the tradition of the Church. As much in interpretation as in origin, scripture is not a disembodied set of writings, but, just as it arose from a community, so it continues to function within the hermeneutical heritage of that community. Indeed, as James Barr has reminded us, the Bible itself is a kind of compressed tradition, as the writings of centuries came to be compiled and contained within one set of covers. Speaking of the Old Testament, he says that 'The effect of scripture was to foreshorten the tradition within the biblical period, to an astonishing degree.'[26] Those who believe in the Holy Spirit will not find it difficult both to acknowledge his work in the development of doctrine and to expect that development to exhibit a continuous relationship with its biblical origins.[27]

Earlier mention of the canon directs me to my second point. It seems to me, as a matter of Christian experience, that one has to strive to take scripture seriously in its totality. An à la carte approach, relying on carefully selected favourite passages, would be an unacceptable impoverishment. I am not suggesting that all parts of the Bible are of equal spiritual stature, that the Epistle of Jude is as valuable as John's gospel, or Esther as important as second Isaiah. There are obviously variations, and there are obviously contradictions and unedifying incidents recorded in its pages. Yet, for reasons which I find hard to articulate more clearly, I feel that I must try to use each part of the scriptures in a way which proves possible and appropriate for it. Only then would the untidiness yet hopefulness of life find its match in the untidiness yet hopefulness of scripture.

[23] Thiselton (1980).
[24] ibid., p. 326.
[25] Quoted in ibid., p. 301.
[26] Barr (1980), p. 50.
[27] It is the apparent lack of anchorage of the Marian dogmas in the New Testament which makes them difficult for me to accept as legitimate developments of doctrine.

The Nicene Creed stops just short of the full articulation of trinitarian doctrine. The three divine Persons are presented to us, but the mysterious subtlety of their interrelationship is not spelled out. Later formulations, such as the *Quicunque Vult* (the so-called Athanasian Creed), attempted to make up the deficit, but they did not attain to the status of ecumenical *symbola*. It was clear both that a theological task remained incomplete and also that its full completion might well elude the power of finite human thought.

Theologians distinguish divine essence (God-in-himself) from divine energy or economy (God as he is known in interaction with his creatures). A bottom-up thinker will find himself concerned with the economic Trinity. The whole thrust of these lectures so far has been to express some understanding of how God is encountered as Father (the transcendent Creator), as Son (made visible in the incarnate life of Jesus Christ) and as Spirit (discreetly at work within the process of our lives and of the cosmos), while holding fast to the fundamental intuition derived from our heritage in Israel, that there is one true God. The same bottom-up thinker can both perceive the value, as well as the necessity, of essaying some further discourse discussing the essential (or immanent) Trinity, and also marvel at the confidence with which many theologians have felt able to speak of the ineffable mystery of the divine nature itself. Faced with talk of *perichorēsis*, and begetting and procession, he is often driven to wonder, 'How do they *know?*' It is far beyond my competence to pursue these matters further, and I must be content with some simple observations.

The first is simply to emphasize that, whatever might be the speculative elaboration of aspects of trinitarian discourse, it had its origin in the Church's struggle to come to terms with its threefold *experience* of divine encounter. The Pauline necessity to speak of God the Father *and* of the Lord Jesus Christ *and* of the Spirit of God and the Spirit of Christ (Rom. 1.7; 8.9) proved persistent. It was the driving force of trinitarian thinking. The proclamation of the One in Three and Three in One is not a piece of mystical arithmetic, but a summary of data.

In science, a good idea excites one by its power to illuminate phenomena other than those which gave rise to its initial conception. Dirac invents an equation to describe the relativistic behaviour of electrons and – lo and behold! – it turns out also to explain a hitherto baffling property of their magnetic behaviour. Such uncovenanted fruitfulness induces the conviction that one is on to something. The doctrine of the Trinity displays a similar illuminating quality for me.

We wish to speak of God as personal, but recognize that he transcends personality as we understand it. It is doubtless crude to appeal to the analogy of the incomprehensibility of a six-faced cube to a dweller in two-dimensional flatland, but a Three-Person God carries some hint of the richness of the transpersonal in a way whose precise articulation is beyond our limited powers of comprehension.

The Eastern Church has often been willing to take seriously a more 'social' picture of the community of the divine Persons within the Trinity than has been accommodated in much Western thought, which has tended to what Moltmann, in rebuke, calls a 'Christian monotheism',[28] minimizing trinitarian insight. A social Trinity delivers us from a static and narcissistic interpretation of the great text 'God is love' (1 John 4.16), and permits a more dynamic account of the divine *agapē*, and so of the divine nature. John Zizioulas states Eastern belief when he writes, 'The being of God is a relational being: without the concept of communion it would not be possible to speak of the being of God . . . It would be unthinkable to speak of the "one God" before speaking of the God who is "communion", that is to say, of the Holy Trinity.'[29] In fairness to the West, Augustine's relational models of the Trinity (particularly: Father = Lover, Son = Beloved, Spirit = the-Love-exchanged) carry a somewhat similar conceptual content of dynamical exchange, though Gunton has criticized Augustine's analogies as being concerned with 'a view of an unknown substance *supporting* the three persons rather than *being constituted* by their relatedness'.[30] The Trinity is not a divine epiphenomenon.

Talk of being-as-communion has a degree of congenial consonance with modern science's way of speaking about the physical world. The advent of both relativity and quantum theory has replaced the Newtonian picture of isolated particles of matter moving along their separate trajectories in the 'container' of absolute space, by something much more interrelational in character, both in its description and behaviour. Both space-and-matter and observer-and-observed display a quality of mutuality. Parts of a quantum system retain a non-local 'togetherness-in-separation', however far apart they may become. Torrance has particularly emphasized this modern generation of 'onto-relational concepts' and compared it with the medieval trinitarian thought of Richard of St Victor.[31]

Jürgen Moltmann has given a trinitarian discussion which centres on Calvary as a divine event: 'what happened on the cross must be understood as an event between God and the Son of God',[32] an event where 'the Son suffers dying, the Father suffers the death of the Son'.[33] In his view, 'Anyone

[28] Moltmann (1981).
[29] Zizioulas (1985), p. 17. See also Gunton (1991).
[30] Gunton (1991), p. 43.
[31] Torrance (1980); (1985).
[32] Moltmann (1974), p. 192.
[33] ibid., p. 243.

who talks of the Trinity talks of the cross of Jesus, and does not speculate in heavenly riddles.'[34] To say that what we can know (the cross) is a reliable guide to what is the case (the mystery of the divine nature) is a kind of theological realism. One thinks of Karl Rahner's celebrated equation of the economic and the essential Trinity.[35] One also thinks of the scientific realist cheerfully proclaiming that 'epistemology models ontology',[36] in an act of commitment to the possibility of apprehending the physical world through our encounter with it. The threefoldness of our religious experience must reflect a threefoldness in the divine nature. From a theological perspective, all forms of realism are divinely underwritten, for God will not mislead us, either in his revelation of himself or in the works of his creation.

However, much of Moltmann's discussion of the divine transaction of the cross seems binitarian in tone; only almost as an afterthought are we told of 'the Spirit which justifies the godless, fills the forsaken with love and even brings the dead alive'.[37] Is this just another example of the Spirit's essential hiddenness? And is that also the reason why Jesus himself seems to have taught so little about the Holy Spirit? The gospels make the Spirit a party to the critical events of Jesus' conception (Matt. 1.20; Luke 1.35) and his baptism and subsequent temptation (Mark 1.9–13, par.), but, apart from the special case of the Johannine discourses, there is little said about him elsewhere.[38] One quite often gets a binitarian feel from reading the New Testament, but the Church's subsequent experience of the Spirit led it to refuse anything less than a trinitarian formulation of doctrine. I have argued elsewhere[39] that theology depends upon the help of 'liturgy-assisted logic'. The doctrine of the Trinity arises from the Christian necessity to say in worship, 'Glory be to the Father, and to the Son, and to the Holy Spirit,' and from the experience of praying in the Spirit through the Son to the Father.[40] Gregory of Nyssa said, 'Concepts create idols. Only wonder understands.'[41]

'We believe in one holy, catholic and apostolic Church'

No scientist could deny the importance of working within the conviviality and tradition of a community, from which he or she has learned the tacit skills of research through an implicit apprenticeship and to whose

[34] ibid., p. 207.
[35] See Ford (1989), pp. 197–8.
[36] cf. Polkinghorne (1989b), ch. 21.
[37] Moltmann, p. 244.
[38] See the discussion in Brown (1985), pp. 169–76.
[39] Polkinghorne (1991), ch. 1.
[40] See Doctrine Commission (1987), ch. 7.
[41] Quoted in Moltmann (1992), p. 73.

judgement mature work is to be submitted for approbation or correction.[42] This is true of the lonely genius, like Henry Cavendish or Albert Einstein, as well as of those of us who belong to the great army of honest toilers. To say that is not to deny that there are times when people of insight rightly stand in lonely defiance of the contemporary viewpoint. An Alfred Wegener, stoutly maintaining his theory of continental drift against currently received opinion, was an *Athanasius contra mundum* of the geophysical world. Yet in the end he was vindicated by the acceptance of his theories by the scientific community, just as in the end Athanasius vanquished the Arians. I have already sufficiently discussed my reasons for believing that in neither case did the eventual settling of the intellectual dust arise from a slothful acquiescence in a socially constructed novel paradigm, brought about by the influence of a superior propaganda machine (chapter 2). There were reasons for the decisions, but reasons which were evaluated within and by a truth-seeking community.

Those who speak of our being in a 'post-modern' era frequently cite as one of its characteristics the recognition that community plays an important role in constituting our being. Philosophically, the notion of person can be discriminated from that of individual by the former adding to the latter's internal states a network of external relationships. Alasdair MacIntyre speaks of moral theory being sought within the *praxis* of a community, of which he takes the rule of St Benedict to be a paradigmatic example.[43]

These considerations provide a contemporary setting hospitable to the idea that the Christian community, the Church, should find a place within the spare confines of the creed. However, simply to regard the Church as the theological equivalent of the communities about which I have been speaking would be to omit an extremely important aspect of its nature, which has been consistently claimed for it by those who are its members. The scientific community looks back to the insights and accumulated knowledge of previous generations and seeks to use that as the platform from which to launch itself into a future of intellectually open discovery. The Church, too, looks back to its foundational events and to the accumulated understandings embodied in its tradition, but it is also orientated towards a future hope in a way which is not paralleled in secular life. In theological terms, the Church is not only a historical community, it is also an eschatological community. This duality of aspect finds expression in a duality of language. The mode of past history finds its articulation in terms of *kerygma* (preaching) and tradition; the mode of eschatological hope finds its articulation in doxology (cf. the great passages of heavenly praise in

[42] Polanyi (1958).
[43] MacIntyre (1981).

Revelation 4; 5.6–14; 7.9–12; etc.). It is particularly the role of the Spirit to create this eschatological orientation in the Church, for, remember, he is the *arrabōn* (anticipatory guarantee) of what is to come (2 Cor. 1.22; 5.5; Eph. 1.14). This is put in startling form by Zizioulas, even in relation to Christ, when he writes:

> Now if *becoming* history is the particularity of the Son in the economy, what is the contribution of the Spirit? . . . [I]t is to liberate the Son and the economy from the bondage of history. If the Son dies on the cross, thus succumbing to the bondage of historical existence, it is the Spirit that raises him from the dead.[44]

A bottom-up thinker will want to ask what is the anchorage in experience which leads the Church to speak of itself not only in historical terms, but also in terms which look beyond history. The answer must be its participation in the Eucharist. Zizioulas says, 'The eucharist *constituted* the Church's being.'[45] Herein is involved Christ and the Spirit; historical remembrance and eschatological hope; *anamnēsis* (present participation in past reality) and *epiklēsis* (the calling-down of the Spirit) and *koinōnia* (a common sharing in Christ's everlasting life). From the first, the Eucharist has had this character of 'already but not yet': 'For as often as you eat this bread and drink the cup, you proclaim the Lord's death until he comes' (1 Cor. 11.26). It is both the commemoration of Calvary and the anticipation of the heavenly banquet of the kingdom of God. I would want to add my own testimony to the mysterious but undeniable experience of the sense of the meeting of past event and future hope in the present reality of the Eucharist.

If so much importance attaches to the Eucharist, one might feel surprise that it receives no formal recognition in the creed. I think that the answer may be that it was so much the acknowledged focus of the Church's spiritual life that its very familiarity meant that it did not require explicit reference. Just as the experience of salvation in Christ was so vivid for the early Church that theories of the atonement were a late development in its thinking, so the real presence of Christ in the sacrament was experienced for centuries before divisive debates about its mode and nature began to perplex the Christian world. In the Christian usage of many centuries, the creed has been recited in the context of the Eucharist. However, that is a development which dates from attempts in the fifth century to eradicate heresy, and originally the creeds were more closely associated with baptism. Yet it seems likely, not least by analogy with the synagogue's great confession of its own faith, the *Shema'*, that Christians gathered to break bread have, from the first,

[44] Zizioulas (1985), p. 130 (cf. Rom. 1.4; 8.11).
[45] ibid., p. 21.

coupled this action with an active confession of their belief in God, and that the great prayer of thanksgiving (*anaphora*) at the heart of the Eucharist has, from as early as we can discern it, expressed the grand themes of creation and redemption.[46]

In my own halting and inadequate attempt to explore a possible understanding of the sacramental real presence of Christ,[47] I was driven to lay emphasis, in accord with much modern sacramental thinking, on the Eucharist as the total action of the gathered company of believers with their gifts of bread and wine laid on the altar. This understanding makes intelligible the presence of Christ and the action of the Spirit, and delivers us from a quasi-magical account of the elements in isolation. I wrote that 'In some manner the bread and wine are an integral part of that whole Eucharistic action in a way neither detachably magical nor dispensably symbolic.'[48] The role of the elements in the eschatological action of the Eucharist is a hint of that cosmic destiny – the redemption of matter as well as humanity – which must be the Creator's purpose for the *whole* of his creation.

A high view of human solidarity is required for a successful Christian ecclesiology, and a corresponding emphasis on the ontological reality of communion. Zizioulas makes such a conception the basis of all his theological thinking, writing in relation to the Trinity that 'it is communion which makes beings "be": nothing exists without it, not even God'.[49] The counterintuitive non-locality, or togetherness-in-separation, found in quantum theory,[50] may be seen as a pale reflection of this fundamental role of interrelationship, visible even at the constituent roots of the physical world.

The classic expression of the corporate character of the Church lies in its being called the body of Christ[51] (Rom. 12.3–8; 1 Cor. 12.12–27; Eph. 4.15–16). The inclusive overtones evoked by the title 'Son of man', find here their full enunciation. Zizioulas puts it with paradoxical force when he writes, 'Christ without His body is not Christ but an individual of the worst type,'[52] which I take to mean that unless there is a way of our incorporation into Christ then we are left in the ultimate hopelessness of our private inadequacies and Jesus becomes for us merely a tantalizingly unattainable ideal. A root requirement of the Christian life is the recognition of our

[46] For a discussion of early Christian worship, see Bradshaw (1992).
[47] Polkinghorne (1989a), pp. 92–4.
[48] ibid., p. 94.
[49] Zizioulas (1985), p. 17.
[50] Polkinghorne (1991), ch. 7.
[51] Robinson (1952).
[52] Zizioulas (1985), p. 182.

heteronomy. We are not able to go it alone, and it is in his service that together we find our perfect freedom.

Once again the Eucharist provides an empirical anchorage for these ideas: 'Because there is one bread, we who are many are one body, for we all partake of the one bread' (1 Cor. 10.17) – words whose force in the liturgy is often sadly diminished by our use of many individual wafers in place of a single common loaf.

One cannot write these words about the high nature of the holy catholic and apostolic Church without being aware of the gap between ideal and reality. In a bitter irony, the Eucharist, theologically understood as being the ground of unity, historically has been the ground of division, as the ecclesiastical overtones resonating in its alternative titles – the Mass, the Holy Communion, the Lord's Supper – only too readily remind us. I personally long not so much for organic union between the Churches, as for the intercommunion of Eucharistic hospitality, so that the Lord's people can gather at the Lord's table without exclusion.[53]

One thinks of the ambiguities of church history – not only the terrible incidents of crusade and inquisition and persecution, but also the fearfulness and pettiness and inadequacies of our lives and of so much of church life. Nietzsche said, 'His disciples will have to look more saved if I am to believe in their Saviour.'[54] Yet against that must be set the continuing movements of reform and renewal, which the believer will attribute to the work of the Spirit. The crusades are also the times of Bernard and the Cistercians, of Francis and the Friars Minor. While the Church is *semper reformanda* (always needing reformation), it is also *semper reformata* (always reformed). It is a company of sinners on the way to salvation. The Church lives in the tension of the already and the not yet, reflecting in its institutional forms the deposit of its historical past, while seeking to be open to the charisms of the Spirit on its pilgrimage to its eschatological future.

The Church is the locus of Christian *praxis*, the exercise of a communal commitment to the life of obedience and witness to the way of Christ. These lectures have been very conceptual in tone, but the religious ideas they are attempting to discuss are not presented to us for our detached perusal. Their acceptance implies consequences for the way we live. We all know that Marx said that the point is to change the world, not simply to interpret it, but the two activities are, in fact, inextricably related. There is a practical circle relating understanding and activity: we must interpret in order to have a discriminating basis for action; we must be prepared to act in order to

---

[53] For a powerful presentation of the Eucharist as the ground and means of Christian unity, see Mackey (1987), pp. 146–56.
[54] Quoted in Taylor (1972), p. 123.

demonstrate commitment to our belief. A novel like Graham Greene's *The Quiet American* shows the dangerous folly of well-intended acts without discernment.

It is the liberation theologians of South America[55] who particularly stress Christian *praxis*. The base communities responding to the gospel in the context of oppression and deprivation are truly bottom-up exponents of Christian understanding. Yet their insight, often linked to a particular kind of revolutionary political programme, cannot be made into a rule for all. There is a great variety of Christian praxis, as St Francis de Sales liked to emphasize, dependent upon its context for its form. What is universal is that the Church, wherever it may be, is called to express its belief in action:

> If a brother or sister is ill-clad and in lack of daily food, and one of you says to them, 'Go in peace, be warmed and filled,' without giving them the things needed for the body, what does it profit? So faith by itself, if it has no works, is dead. (Jas. 2.15–17)

[55] Gutiérrez (1988); Rowland and Corner (1990).

9

Eschatology

'We look for the resurrection of the dead,
and the life of the world to come'

Cosmologists do not only peer into the past. They can attempt to discern the future. On a cosmic scale, the history of the universe is a gigantic tug-of-war between the expansive force of the big bang, driving the galaxies apart, and the contractive force of gravity, pulling them together. These two effects are so evenly balanced that we cannot tell which will win. Accordingly, two alternative scenarios must be considered. If expansion prevails, the galaxies now receding from each other will continue to do so for ever. Within each galaxy, gravity will bring about condensation into enormous black holes, which will eventually decay into low-grade radiation through a variety of possible physical processes. On this scenario, the universe ends in a whimper. If, on the other hand, gravity prevails, the present expansion of the galaxies will one day be halted and reversed. What began with the big bang will end in the big crunch, as the whole universe collapses back into a singular cosmic melting pot. Neither of these catastrophes will happen tomorrow; they lie tens of billions of years into the future. Nevertheless, one way or the other, the universe is condemned to ultimate futility, and humanity will prove to have been a transient episode in its history.

It may seem a melancholy prospect, a clear denial of those religious assertions of a purpose at work within cosmic history. Bertrand Russell once wrote that 'Only within the scaffolding of these truths, only on the firm foundation of unyielding despair, can the soul's habitation henceforth be safely built.'[1] I accept the first of his provisos, but not the second. The bleak prognosis for the universe puts in question any notion of evolutionary optimism, of a satisfactory fulfilment solely within the confines of the unfolding of present physical process. That is why many of us find the tone of the writings of Teilhard de Chardin[2] to be ultimately unhelpful. But

[1] Quoted in Peacocke (1979), pp. 89–90.
[2] Teilhard de Chardin (1959).

Christian theology has never staked its claims on the basis of an evolutionary optimism, of the coming of the kingdom of God simply through the flux of history. One is surprised to read John Macquarrie's comment, 'Let me say frankly, however, that if it were shown that the universe is indeed headed for an all-enveloping death, then this might seem to constitute a state of affairs so negative that it might be held to falsify Christian faith and abolish Christian hope.'[3] The answer to that strange remark is found in the words of the creed quoted at the beginning of this chapter. An ultimate hope will have to rest in an ultimate reality, that is to say, in the eternal God himself, and not in his creation.

I do not think that the eventual futility of the universe, over a timescale of tens of billions of years, is very different in the theological problems that it poses, from the eventual futility of ourselves, over a timescale of tens of years. Cosmic death and human death pose equivalent questions of what is God's intention for his creation. What is at issue is the faithfulness of God, the everlasting seriousness with which he regards his creatures. When Jesus was in argument with the Sadducees about the resurrection of the dead, this was the point to which he appealed: 'As for the dead being raised, have you not read in the book of Moses, in the passage about the bush, how God said to him, "I am the God of Abraham, and the God of Isaac, and the God of Jacob"? He is not God of the dead, but of the living' (Mark 12.26–7, par.). In other words, the faithful God is not one who abandons the patriarchs once they have served his purpose, but he has an eternal destiny for them.

How credible is such a hope for humanity? I have already (chapter 1) explained that my understanding of our nature is not framed in the dualist terms of an incarnated soul. The Christian hope is, therefore, for me not the hope of *survival* of death, the persistence *post mortem* of a spiritual component which possesses, or has been granted, an intrinsic immortality. Rather, the Christian hope is of death and *resurrection*. My understanding of the soul is that it is the almost infinitely complex, dynamic, information-bearing pattern, carried at any instant by the matter of my animated body and continuously developing throughout all the constituent changes of my bodily make-up during the course of my earthly life. That psychosomatic unity is dissolved at death by the decay of my body, but I believe it is a perfectly coherent hope that the pattern that is me will be remembered by God and its instantiation will be recreated by him when he reconstitutes me in a new environment of his choosing. That will be his eschatological act of resurrection. Thus, death is a real end but not the final end, for only God himself is ultimate. Although there have, of course, been strands of the Christian tradition which have used the language of the survival of an

[3] Macquarrie (1977), p. 256.

immortal soul, I believe that the tradition which is truer, both to New Testament insight and to modern understanding, is that which relies on the hope of a resurrection beyond death.

If this psychosomatic understanding is correct, then it is intrinsic to true humanity that we should be embodied. We are not apprentice angels, awaiting to be disencumbered of our fleshly habitation. Our hope is of the resurrection of the *body*. By that I do not mean the resuscitation of our present structure, the quaint medieval notion of the reassembling of bones and dust. In a very crude and inadequate analogy, the software running on our present hardware will be transferred to the hardware of the world to come. And where will that eschatological hardware come from? Surely the 'matter' of the world to come must be the transformed matter of this world. God will no more abandon the universe than he will abandon us. Hence the importance to theology of the empty tomb, with its message that the Lord's risen and glorified body is the transmutation of his dead body. The resurrection of Jesus is the beginning within history of a process whose fulfilment lies beyond history, in which the destiny of humanity and the destiny of the universe are together to find their fulfilment in a liberation from decay and futility (cf. Rom. 8.18–25).

The picture of such a cosmic redemption, in which a resurrected humanity will participate, is both immensely thrilling and deeply mysterious. Yet its unimaginable future has a present anchorage in our hearts. There is a deep-seated intuition of hope, all the ambiguities and bitternesses of history notwithstanding, which encourages the belief that in the end all will be well. This is so widely diffused as to constitute, by its presence, a 'signal of transcendence', even among those who might claim no explicit religious belief.[4] It is important that the Christian Church does not lose its nerve in witnessing to the coherence and divine assurance of such a hope. I like the words of John Robinson when he says that for the Christian, he or she 'knows the present for what it is; that is to say, a point too charged with eternity to be understood except by myths which open a door into heaven and force upon every moment the terrible relevancy of the first things and the last, the elemental and the ultimate'.[5] Needless to say, our recourse to mythic language is in order to speak of that which is not yet experienced, and not in the form of the illusory comfort of a fable.

Some have wondered whether, despite the eventual bleakness of the cosmic future, there might not be some form of satisfactory fulfilment possible within its process. Humanity, and carbon-based life generally, will certainly disappear, but might not 'intelligence' be able to engineer further

[4] Berger (1970), pp. 72–6.
[5] Robinson (1950), p. 70.

embodiments of itself, appropriate to changing cosmic circumstances and permitting either its infinite continuance within the decaying phase of an expanding universe, or the processing of an infinite amount of information during the hectic, highly energetic, final moments of a collapsing cosmos?[6] Such a 'physical eschatology' has been pursued with the greatest persistence and ingenuity by Frank Tipler.[7] It is the collapsing case which presents, he believes, the possibility of a sequence of ever faster-racing cosmic computers, whose asymptotically infinite capacity would make it appropriate to speak of the final instant of the present universe as culminating in the fleeting achievement of the Omega Point, the physical realization of a cosmically evolved 'god'. In the closing phases of this scenario, Tipler envisages the enormous computing power available being used to construct computer simulations of ourselves (we being understood by him as constituting finite-state machines), so that even the hope of 'resurrection' finds a place within his physical eschatology!

It is a fantastic, and curiously chilling, programme. I do not think that it succeeds. There are physical difficulties: the unlimited amounts of computer power are attained only in the infinitesimal dying instants of the universe, and their availability depends upon taking absolutely literally extrapolations into this unknown area of physical process. Highly conjectural properties of matter would be required. Tipler's speculations about the very last moments of cosmic history make the speculations of other quantum cosmologists about the very earliest instants of cosmic history look quite tame by comparison. There are anthropological difficulties: I have already explained that I do not think that we are 'computers made of meat' (chapter 1), so that the finite-state machine model of humanity is unacceptable. There are teleological difficulties: the ambiguous history of intelligence to date does not wholly encourage the view that these vast amounts of computing power would be used for such 'benevolent' purposes. There are also intuitive difficulties. They are hard to articulate, but the scheme has far too abstract a feel to it to represent the satisfaction of a truly human hope. In religious terms, it corresponds to a kind of cosmic tower of Babel, the fundamental error of confounding creation with its Creator. I regard physical eschatology as presenting us with the ultimate *reductio ad absurdum* of a merely evolutionary optimism.

I think that Tipler's answer to some of these criticisms would lie in his concept of what he calls the Teilhard Boundary Condition, claiming to *specify* the attainment of that final Omega Point. He uses it, within a many-worlds quantum cosmology, in a way analogous to Hawking and Hartle's

---

[6] Barrow and Tipler (1986), ch. 10; Dyson (1988), ch. 6.
[7] Tipler in Russell et al. (1988), pp. 313–32.

employment of a 'no boundary' condition in their speculative version of cosmology. Since the cosmic Schrödinger equation in principle determines the wavefunction of the universe once a boundary condition has been imposed, this then appears to guarantee Tipler's desired scenario, and he can use language such as 'the ultimate future guides all presents into itself'.[8] There is at least some sort of verbal parallel with Pannenberg's eschatological emphasis on the power of the future, and Pannenberg himself seems to accord some significance to this.[9] Yet, as Drees in a helpful discussion has clearly emphasized,[10] Tipler's reliance upon a reductionist and essentially determinist[11] view of quantum cosmology makes his thinking much closer to a kind of Spinozan pantheism than any recognizable form of Christian theology. His 'god's' claimed transcendence is the peripherality of a boundary condition, not the eternal independent freedom of the Ground of hope. The specified ultimate future is able to exercise its influence on the present precisely because the future is not open in Tipler's deterministic quantum universe. I remain, therefore, unconvinced by these fantastic speculations.

If that is right, we are driven back, as we might have expected, to God alone as the basis of final hope, so that our own and the universe's destiny awaits a transforming act of divine redemption. In Christian thought this is expressed in terms of a new creation (2 Cor. 5.17), a new heaven and a new earth (Rev. 21.1–4). The resurrection of Christ within history is then understood as the anticipation of this great event lying beyond history, the seed from which eschatological fulfilment will eventually blossom for all (1 Cor. 15.20–8). There is an obvious difficulty in such a conception, which one might characterize as the pie-in-the-sky objection, written cosmically large. Does the future hope not devalue the present reality, by making the former the true existence and the latter only an unsatisfactory prelude to it? Indeed, one might add, an unnecessary prelude, for if the new creation is going to be so wonderful – and its nature is expressed in terms of a picture where 'death shall be no more, neither shall there be mourning nor crying nor pain any more, for the former things have passed away' (Rev. 21.4) – why did God bother with the old? The 'matter' of that world-to-come must be such that it will not enforce recapitulation of the deadly raggednesses and malfunctions of the present universe, otherwise the prospect would be of a

[8] *Zygon* 24, p. 240.
[9] ibid., pp. 255–71.
[10] Drees (1990), pp. 128–41.
[11] Many-worlds quantum theory is based *solely* on the Schrödinger equation and so it is intrinsically determinist in character. In quantum cosmology of this kind, time is a *secondary* construct in what is a world of polyvalent being, rather than a world of true becoming.

kind of eternal return, an endless rerun of this vale of tears. (The rejection of that repetition is what I take to lie behind the words of Mark 12.25, par.) Yet, if the natural laws of the new creation are such as to permit such a redemption of embodied existence, why were they not the basis for the first creation? Any realistic account of Christian eschatology must take these questions seriously.

I think the answer lies in the recognition that the new creation is not a second attempt by God at what he had first tried to do in the old creation.[12] It is a different kind of divine action altogether, and the difference may be summarized by saying that the first creation was *ex nihilo* while the new creation will be *ex vetere*. In other words, the old creation is God's bringing into being a universe which is free to exist 'on its own', in the ontological space made available to it by the divine kenotic act of allowing the existence of something wholly other;[13] the new creation is the divine redemption of the old. Concerning the new creation, Gabriel Daly says, 'The word "new" could mislead here. It does not imply the abolition of the old but rather its transformation. It is a "new creation" but, unlike the first creation, it is not *ex nihilo*. The new creation is what the Spirit of God does to the first creation.'[14] The understanding that this creates in my mind is that the old creation has the character which is appropriate to an evolutionary universe, endowed with the ability through the shuffling explorations of its happenstance to 'make itself'. It is a universe, certainly not lying outside the sustaining and providential care of God, but nevertheless it is given its due independence to follow its own history. That historical process cannot avoid the cost of suffering which is the price of independence. The new creation represents the transformation of that universe when it enters freely into a new and closer relationship with its Creator, so that it becomes a totally sacramental world, suffused with the divine presence. Its process can be free from suffering, for it is conceivable that the divinely ordained laws of nature appropriate to a world making itself through its own evolving history should give way to a differently constituted form of 'matter', appropriate to a universe 'freely returned' from independence to an existence of integration with its Creator. It is, of course, far easier to see what the phrase 'freely returned' might mean in relation to humanity than it is in relation to the whole universe. Nevertheless, the intuition that our destiny is intimately bound up with the destiny of the cosmic womb from which we were born forces one to try to grope for that wider meaning.

There are some clues to be found in the tradition. The incarnation

[12] Polkinghorne (1991), ch. 8.
[13] Moltmann (1981), pp. 105–14; (1985), ch. 4.
[14] Daly (1988), p. 100.

established a bridge between Creator and creature, and it is theologically natural to look to the cosmic Christ, in whom 'all the fullness of God was pleased to dwell, and through him to reconcile to himself *all things*, whether on earth or in heaven' (Col. 1.19–20), for that path of universal return. The Eastern Orthodox notion that the ultimate destiny of the whole creation is *theosis* (deification) points us in a similar direction.[15] One might say that panentheism is true as an eschatological fulfilment, not a present reality. I have already spoken more than once of the way the empty tomb testifies to an abiding future for matter; and the hints of continuity and discontinuity (sharing of food; appearance and disappearance within closed rooms) which the gospel accounts of the resurrection appearances give us can be interpreted as indications appearing within history of the transformed nature of eschatological 'matter'. Paul wrestles with such problems when he speaks of the relationship between the present physical body and the future spiritual body (1 Cor. 15.35–57). Because the Church is an eschatological community as well as a historical community (chapter 8), its present experience might be expected at least to contain some type or shadow of things to come. Whatever may be the mode of Christ's presence in the Eucharist, it certainly involves some link up of the present with the future of the risen Lord. Gerald O'Collins says, 'The Eucharist provides the supreme instance of how the resurrection has already changed the created world . . . What happens here points ahead to the final state of the material universe when it will be publicly and fully under the risen power of the Lord.'[16]

Great mysteries remain, which only experience of the *eschaton* itself will begin to remove from us. Although I am incapable of illuminating the root of the matter further, I want to go on to ask what understanding might result from this picture of the new creation as arising from the transformation of the old when it is applied to the traditional Christian concerns relating to creaturely destiny.

One immediate consequence of the new creation being *ex vetere* is that, quite contrary to the jibe about pie-in-the-sky, it invests the present created order with a most profound significance, for it is the raw material from which the new will come. The abolition of a mere evolutionary optimism in no way abrogates a present concern, neither in relation to ecological responsibility for a just and sustainable enjoyment of nature, nor in relation to the effects of our individual decisions upon the development of our characters. What is to be will come from what is presently the case. That is so, not only in relation to the parochial concerns of terrestrial history, but also in relation to the grand sweep of the development of the universe. God

[15] Lossky (1957), ch. 5.
[16] O'Collins (1987), p. 156.

may be doing all sorts of unexpected things throughout the vastness of the cosmos. His is not the tight control of the divine puppet-master, pulling every cosmic string, but his gift of a genuine independence to the creation has not made him into an indifferent or impotent spectator of its process. He is in continual providential interaction with his world, and it may well be that within its history he will bring about purposed ends, even if they are achieved along contingent paths.[17] The metaphor of God as the Grand Master of cosmic chess, responding to, but never baffled by, the moves of his creaturely 'opponent', and so sure to deliver eventual checkmate as the game unfolds, is one which has been popular with many theological writers ever since William James first made use of it.

The old creation has its own fruitfulness and brings about its own possibilities. Yet it must be delivered from the frustration of its impending mortality, just as Jesus was delivered from the bonds of death by his resurrection. In each case a great act of God is called for, but an act which must be the fitting fulfilment of what has gone before, not its arbitrary abolition. Just as the cross and the resurrection are part of the one drama of the incarnation, so the old and new creation must be part of the one drama of God's purpose for his creatures. What we must be at pains wholly to exclude is any magical notion of a divine *tour de force* simply putting right, through the exercise of naked power, something which had otherwise got out of control. That is simply theologically unacceptable, on the general grounds of what a personal God must be like, ever respectful of that to which he has given the gift of freedom. It is also unacceptable to a bottom-up thinker, who reads from the long history of cosmic evolution the story of a God of process and not of magic, a God who is patient and subtle in the achievement of his purposes. Yet such a God must not simply be in thrall to the eventual futility of universal or individual history. Moltmann says, 'Death can set no limits to the unconditional and hence universal love of God . . . Otherwise God would not be God and death would have to be called an anti-God.'[18] The concept of a new creation *ex vetere* is the attempt to do justice both to the God of process and to the God of hope. It emphasizes that the One who creates is also the One who redeems.

That hope of a transformed new creation in which all will participate is the only ultimate answer to the surd of suffering present in the old creation, for theodicy cannot be based on the idea that the agony of past generations is the necessary price of some future evolved happiness for others alone. Each generation must receive the healing and fulfilment that is its due. Moltmann says of the sufferings of every age that 'They do not acquire any meaning in

[17] See Bartholomew (1984), ch. 4.
[18] Moltmann (1990), p. 190.

history, not even through future history; they can only await their redemption.'[19] Speaking of the necessity that 'each sentient being must have the opportunity of achieving an overwhelming good', Ward says bluntly, 'theism would be falsified if physical death was the end, for then there could be no justification for the existence of the world'.[20] He feels this so strongly that he argues for some form of continuing destiny for the higher animals as well as for humans.

Eschatology is indispensable to theology. 'Christ's parousia is not a dispensable appendage to the history of Christ. It is the goal of that history, for it is its completion.'[21] 'For he must reign until he has put all his enemies under his feet. The last enemy to be destroyed is death . . . When all things are subjected to him, then the Son himself will also be subjected to him who put all things under him, that God may be everything to every one' (1 Cor. 15.25–6, 28).

More detailed thoughts on how the old and the new creations relate to each other are necessarily extremely speculative. Because modern physical understanding associates matter and spacetime intimately with each other, it is a natural to suppose that the 'time' of the new creation bears some sequential relationship to the time of the old – it comes 'after' or 'beyond' that transformation of matter into 'matter'. On the other hand, the considerations relating to the resurrection appearances, and even to the Eucharist, might encourage the thought of some degree of intersection between the historical and the eschatological. (See the discussion on pp. 121–2.) What does seem clear is that if it is intrinsic to humanity to be embodied, then it must be intrinsic to humanity to be temporal. The life of the world to come will doubtless be everlasting, but it will not be eternal in that special and mysterious timeless sense in which the word is applied to God himself. The patient process of this world will find its reflection in the redemptive process of the world to come. Our notion of heaven is delivered from any static, and potentially boring, conception. The life of heaven will involve the endless, dynamic exploration of the inexhaustible riches of the divine nature.

Involved in that redemptive process will be judgement and purgation. The stern images of the New Testament (Matt. 25.31–46; Rom. 2; Rev. 21.5–8; etc.) which point in this direction are to be taken with all seriousness. Yet, I think we must read them aright, seeing that there is both light and darkness, something of the sheep and something of the goat, within each one of us. The sinful dross of our lives must be refined away. The process will be painful as we are brought face to face with the reality of what

---

[19] ibid., p. 207.
[20] Ward (1990), p. 201.
[21] Moltmann (1990), p. 316.

we are, but in essence it is hopeful. Daly says of sin that it is 'ultimately a hopeful word'[22] for it points to the possibility of forgiveness, and Robinson says of judgement that its 'sole possible function . . . can be to enable men to receive the mercy which renders it superfluous'.[23] The controlling divine image is that of the healing surgeon, not the wrathful potentate. There is an important need for Christian theology to recover a credible, demythologized concept of purgatory if it is to be able to speak of the merciful God who works through process and not by magic.

If what I have said is true, it follows that there cannot be a kind of curtain which comes down at death, dividing humanity irreversibly into the companies of the saved and of the damned. God's loving offer of mercy cannot be for the term of our earthly life alone, so that it is withdrawn after three score years and ten. Here is the answer to the perplexity of what is to be the destiny of those who for one reason or another have never truly been faced with the claims and the promise of the gospel. To say that is not to diminish our responsibility to witness to that gospel, for that would follow simply from our obligation to testify to the truth as we have found it. Nor is it to diminish the significance of the decisions we make in this life. Every turning away from God will make the return journey that much the harder. A flippant indifference, such as was expressed by the poet Heine on his deathbed, when he said, 'Dieu me pardonnera. C'est son métier' (God will pardon me. It's his line of business), is spiritually damaging. We are not to presume upon divine mercy.

But will there be those who will refuse the divine pardon and purgation for ever, or in the end will all be saved (even Hitler and Stalin)? It is well known that the New Testament seems sometimes to speak in universalist terms (e.g., Rom. 11.32; 1 Cor. 15.22) and sometimes in terms of some who will be lost (e.g., Matt. 25.46; Rom. 2.6–11). I cannot believe that God will ever foreclose on his loving offer of mercy, but equally I do not believe he will override the human freedom to refuse. If there is a hell, its doors, as the preachers say, are locked on the inside. Those who are there are there by choice. It is not a place of torment, but rather a place of exquisite boredom, for it has all the emptiness of life without God. This was imaginatively portrayed for us by C. S. Lewis in his picture of hell as a grey drab world lost down a crack in the floor of heaven.[24]

If these ideas are correct, they illustrate the claim that theology can make to be a discipline concerned with the progressive exploration of truth, not held for ever in thrall to past understanding alone. One could not deny that

[22] Daly (1988), p. 128.
[23] Robinson (1950), p. 106.
[24] Lewis (1946).

the terrors of hell (so chillingly portrayed in Dante's *Inferno* and Bosch's paintings) were for many centuries standard Christian doctrine and that they were powerfully invoked by preachers to encourage repentance and to enforce discipline. Nor could one deny that there are passages in the New Testament which, taken at their literal face-value, would seem to support this view. It was in the nineteenth century that serious questioning of this interpretation began, and it led in the first instance to some sad heresy trials in the churches. Eternal punishment was a source of moral scandal which helped to alienate many thoughtful and sensitive people from contemporary Christianity. Charles Darwin called it a 'damnable doctrine' and said he could not 'see how anyone ought to wish Christianity to be true'.[25] It has now largely been abandoned. This has come about, not through surrender to a secular sentimentality, but through the realization of its incompatibility with the mercy of a loving God, who cannot be conceived to exact infinite punishment for finite wrong. Theology has proved itself to be open to correction.

Our belief in a destiny beyond death rests ultimately on our belief that God is faithful and that he will not allow anything of good to be lost. Our understanding of the value of individual human life in his sight (the God of Abraham, Isaac and Jacob, the God of individuals) leads us to expect a personal destiny for each one of us. It is not sufficient for Christian understanding to think simply of a return of the many to the One, the reabsorption of the drops of our being in the ocean of divine Being. That would produce, as Zizioulas says, 'a blessedness in which there would be no blessed'.[26] Because human relationships are part of the good that we presently experience (as well as being in need of redemptive healing), one must rightly expect their restoration and fulfilment in the world to come. A parish priest is frequently asked by the bereaved whether they will meet their loved ones again beyond the grave. I think that trust in a faithful and loving God enables one to give that assurance.

The good that is not to be lost is not confined to human good alone. God must have an eschatological purpose for all of this creation. Yet we assign value to that creation in different ways that seem to us appropriate to the nature of that with which we have to do. I think we should trust those intuitions. With the animals, we attach significance rather more to the species than to the individual. We would be sorry if all deer were exterminated, but the culling of individuals to preserve the herd seems to most of us to be an ethically acceptable practice. I suspect that the destiny of much of creation will be in this typical, rather than individual, mode. (I leave

---

[25] Desmond and Moore (1991), p. 623.
[26] Zizioulas (1985), p. 166.

aside the intriguing question of pets who have acquired individual value through their close association with their human owners.)

How will we experience our eschatological destiny? Will we make an immediate transition at death to the 'time' and life of that world to come, so that death and resurrection follow hard upon each other, or will we (as most traditional Christian thought has held) be kept in some intermediate heavenly 'holding pattern', awaiting the final resurrection of the dead and the consummation of all things? Perhaps the best answer to such a question is, 'Wait and see.' Nevertheless, we may permit ourselves a few speculative thoughts. Either kind of scenario seems a coherent possibility, and the latter finds an experiential anchorage in the widespread Christian intuition that it is good to remember the departed in our prayers. We are such timebound creatures that our minds can easily balk at thinking how time and 'time' might relate to each other, but here is one of the few points at which the mathematically minded may have a theological advantage, for they are routinely able to conceive how such relationships might be expressed. I do not want to labour the point, but the imagery of vector spaces orthogonal to each other (the old and new creations) but connected by mappings or projections (resurrection, redemptive transformation) or even intersections (resurrection appearances) affords a conceptual framework within which such notions could be contained.

In terms of the general stance adopted in these lectures, the traditional idea of an intermediate state between death and the End could be accommodated, and it would find its natural expression in terms of those remembered patterns of ourselves held in the mind of God (the preserved 'software' awaiting a new realization through resurrection and, perhaps, subject to some 'debugging'). Because it is of the essence of humanity to be embodied, such a state of remembrance would be less than fully human, very much the equivalent in modern terms of the Hebrew notion of shades in Sheol, a 'life on half pay' in Robinson's graphic phrase.[27] For the process theologians, it seems that this etiolated state is the sum total of their hope. Pailin tells us that 'the proper goal of divine creativity is not to be envisaged as the attainment of a particular state of affairs but as the continual pursuit of aesthetic enrichment',[28] that divine enrichment arising in part from the deposit that our lives make in the reservoir of God's memory. Traditional Christian understanding offers us a much more fulfilling prospect than contributing to the building up of a kind of spiritual coral reef.

A final problem remains. How can that future life of finite creatures be proof against the peril of a further fall back into renewed rebellion and

[27] Robinson (1950), p. 76.
[28] Pailin (1989), p. 132.

sinfulness? After all, if God declined to populate his first creation with perfect automata, he will surely not do so in the new creation either. Yet how then can that new creation be delivered from a perpetual precariousness? Aquinas identified freedom with consent to the good, and in the clear light of the beatific vision I think we may expect to be able fully and joyfully to commit ourselves to him whose service is perfect freedom. Once again, an insight of process theology finds its true application eschatologically. In the unveiled openness of the new creation to its Creator, divine persuasion will indeed be that which brings about the divine will, without violence to the freedom of the creature.

Eschatology might seem a queer subject for a self-confessed bottom-up thinker to address. Yet I think that to do so is an essential exercise, for it does not involve mere ungrounded speculation about what might lie hidden in the mists beyond the end of history, but rather it essays the articulation of what is the coherent ground of present hope. We must have the intellectual boldness to defend that hope. 'If in this life only we have hoped in Christ, we are of all men most to be pitied' (1 Cor. 15.19, RSV margin).

Eschatology also reminds us that religious belief involves truth claims of unusual kinds, relating to experience currently inaccessible to us but open to confirmation in the future. From this arises John Hick's notion of eschatological verification.[29] He speaks of the parable of two men on a journey. One believes that they are going to the celestial city, the other has no such belief. On the way their experiences are closely similar. 'And yet when they do turn the last corner it will be apparent that one of them has been right all the time and the other wrong.'[30] In the End, I believe that we shall see the visible vindication of the God with whom, in this present life, we can only walk by faith. Yet also, those who 'make it clear that they are seeking a homeland . . . [who] desire a better country, that is, a heavenly one' (Heb. 11.14, 16), will tend to conduct themselves in the present in a manner different from those without such a hope, so that even on the way there will be a difference between the two travellers. The one will endure privations for the sake of the prospect before him; the other will be tempted to stray on to paths of ease, even if they deflect him from his journey. It will be easier for the one than for the other to honour the sometimes painful obligations that loyalty and faithfulness impose upon us.

The point being made in talk of eschatological verification is epistemological; our knowledge will not be complete without that *post mortem* experience when faith will vanish into sight. Yet Pannenberg, in his emphasis on the constitutive role of the future, seems to go further and to accord

[29] Hick (1966a), ch. 8.
[30] ibid., p. 177.

eschatology an ontological significance. Grenz says, 'He boldly declares that in a sense God does not yet exist, because God's existence and deity are bound together in God's future kingdom. There he explicitly speaks ontologically and not merely epistemologically . . .'[31] Of course, Pannenberg partly takes that back with his talk of the retroactive influence of the future on the present, but it seems to me better to avoid the contortions of thought imposed by so great an emphasis on the End, and simply to recognize that the Ground of future hope is present faithful Reality.

[31] Grenz (1990), p. 74.

10

Alternatives

The Nicene Creed was formulated in the course of the same century that had earlier seen Constantine's conversion, with its consequence that, for a long while after, the theological debate was internal to Christianity. The 'many "gods" and many "lords" ' (1 Cor. 8.5) of the Mediterranean world disappeared, as would the gods of Northern Europe, while the rift with Judaism was too deep for serious exchange to take place between the two religions for many centuries. For several centuries after the rise of Islam, the principal Christian response to this new religion was by way of resistance to its incursions and attempts at reconquest. How different is the situation today! World-wide communications, and extensive immigrations, have made us only too aware that Christianity is but one among the several great historic traditions present in the world of the faiths. For a bottom-up thinker there is a perplexing contrast with the spread of modern science. Originally the product of Western Europe, it has proved eminently exportable, so that one can expect to receive the same answer to a scientific inquiry, whether it is made in London or Tokyo, New York or Delhi. In contrast, while there is some degree of Christian presence in almost every country, in many it is tiny and the other historic religious traditions have shown great stability in the face of more than two centuries of widespread Christian missionary effort. It is a pressing problem for a credible theology, second only to the problem of suffering, to give some satisfactory account of why the diversity of religious affirmations should not lead us to the conclusion that they are merely the expressions of culturally determined opinions. Kenneth Cragg[1] reminds us that even in the seventeenth century John Bunyan felt the difficulty. In *Grace Abounding* he wrote, 'Everyone doth think his own religion rightest, both Jews and Moors and Pagans: and how if our faith, and Christ, and scriptures, should be but a think so too?'

Of course, there is unquestionably a degree of cultural determination in our actual religious beliefs. If I had grown up in Saudi Arabia, rather than in England, it would be foolish to deny that the chances are I would be a

---

[1] Cragg (1986), p. 315.

Muslim. But the chances are also that I would not have spent most of my life as a theoretical physicist, but that does not mean that science is simply a cultural artefact.[2] We must not commit the genetic fallacy of supposing that origin explains away the content of belief.

To some extent the effect of culture is the inescapable deposit of the separate historical developments of communities. For Pannenberg this appears to be a sufficient insight: 'the plurality of viewpoints in the struggle for the one truth – a plurality which is insuperable within the historic process – becomes intelligible because of the openness of history'.[3] That does not seem to me to be enough. As with the problem of suffering, the difficulty lies not in the existence of the phenomenon, but in its scale. That there should be diversities of religious understanding is not surprising; that the discrepancies in the accounts of ultimate reality are so great, is very troubling. That perplexity is increased when we consider that it is knowledge of *God*, with all his power to make himself known, which we are considering. An American Indian said to a missionary, 'If this faith is so true why was it not given to *our* ancestors?'[4] (Cragg points out that an Englishman could reply that it wasn't given originally to his ancestors either. Some propagation of locally given revelation through space and time is not an incoherent possibility for personal divine action.)

There have been three broad avenues of approach to the problem of religious diversity.[5] The approach which is usually called pluralism regards the world's religious traditions as being, in essence, equally valid expressions of the same fundamental religious quest, different pathways up the spiritual mountain. Its driving force is the conviction that God cannot have left himself without witness at most times and in most places; that most people cannot have been cut off from his saving grace just by the accidents of circumstance. One of its chief proponents is John Hick, who writes, 'Can we then accept the conclusion that the God of love who seeks to save all mankind has nevertheless ordained that men must be saved in such a way that only a small number can in fact receive salvation?'[6] I have already made it clear (chapter 9) that I agree with him in answering 'No' to that question. But ultimate universal access to salvation does not require the proposition of the essentially equal validity of all current religious points of view. Hick's pluralist strategy is based on viewing religious traditions as alternative

---

[2] I know that some claim the contrary; see Polkinghorne (1989b), ch. 21, for my response.

[3] Pannenberg (1976), p. 421.

[4] Cragg (1986), p. 181.

[5] D'Costa (1986); Race (1985). For a critique of this categorization, see Barnes (1989).

[6] Quoted in D'Costa (1986), p. 14.

schemes of salvation, means for 'the transformation from self-centredness to Reality-centredness'.[7] The Real itself is inaccessible, and it is only the culturally formed personal or impersonal masks of Reality which the world faiths present to us. Hick's strongly instrumentalist view of religion means that, for the traditions, 'their truthfulness is the practical truthfulness which consists in guiding us aright'.[8] 'The basic criterion, then, for judging religious phenomena is soteriological.'[9] No one should deny the import-ance of religious *praxis* – 'the tree is known by its fruit' (Matt. 12.33, par.) – nor the presence of compassion in all the traditions, but a purely pragmatic account is as unsatisfactory for religion as it is for science.

When we come later to consider some of the conflicts of understanding between the traditions, we shall see how difficult a pluralist position is if one wishes (as I do) to assign cognitive, rather than merely expressive or dispositional, content to religion. Commenting on Hick's programme claiming to discern a noumenal common denominator, Ward says that 'The assertion that "only the vague is really true" seems highly dubious; but even if it is made, one is making a selection from a wider range of competing truths in religion.'[10] It is just not the case that, under the skin, the world's religions are really all saying the same thing, and one can question whether the attempt to impose pluralism on the traditions does not lead, as Schwöbel says, 'to a personal construction of the history of religions and religious attitudes that very few who participate in them would recognize as their own'.[11] The driving force of much pluralist thought is the desire to iron out differences in the search for tolerance, but this 'can all too easily turn into a new guise of Western imperialism where subscribing to the principles of the Enlightenment becomes a precondition for participation in dialogue'.[12] The particularities of the traditions must be respected, for to ask an adherent to give up what are perceived as the central truths of a tradition in order to find accommodation within a sufficiently unfocused universality is 'akin to asking a native speaker of English to please try and do without nouns, since we have reason to believe that using them leads to an inappropriately reified view of the world'.[13]

But neither, it seems to me, can we take the second option of an exclusivist approach and say that the truth is with us and the others are just mistaken. Our contacts with the other traditions make clear to us that they

[7] Hick (1989), p. 36.
[8] ibid., p. 375.
[9] ibid., p. 309.
[10] Ward (1990), p. 199.
[11] D'Costa (1990), p. 42.
[12] ibid., p. 33.
[13] P. J. Griffiths, in ibid., p. 168.

are the vehicles of profound spiritual experience and understanding. I recall a television programme in which Ronald Eyre interviewed a Buddhist Zen master. I suppose Zen is about as far away as could be from a religious understanding with which I feel at home, but the authenticity of that person's spiritual life was vividly conveyed, even through the medium of the box. Even someone like Hendrik Kraemer, who took a sternly exclusive view of the supremacy of Christianity, acknowledged that 'undeniably God works and has worked in men outside the sphere of biblical revelation'.[14] The old formula, strictly interpreted, '*extra ecclesiam nulla salus*' (no salvation outside the Church), just will not do, as Vatican II came to recognize. Yet if, as these lectures maintain, Jesus Christ is the unique meeting-point between human need and the life of God, then, in an ultimate sense, there is an exclusivism in Christ (cf. John 14.6). Moltmann wrote, 'Outside Christ no salvation. Christ has come and was sacrificed for the reconciliation of the whole world. No one is excluded.'[15] But more know him than know him by name (cf. John 1.9).

That is how I would wish to interpret Karl Rahner's celebrated phrase for adherents of other faiths: 'anonymous Christians'. It has sometimes been seen as a patronizing or imperialistic way of speaking, but the basic concept, rightly construed as the implicit presence of the Word and the covert activity of the Spirit throughout the breadth of the human religious quest, can be freed from those undesirable overtones. It then serves as an expression of the third approach, that of inclusivism. D'Costa defines this as 'one that affirms the salvific presence of God in non-Christian religions while still maintaining that Christ is the definitive and authoritative revelation of God'.[16] That is certainly the stance that I myself would wish to adopt. It has a venerable tradition in Christian thought. Justin Martyr in the second century, and Clement and Origen in the third, invented the category of 'good pagans', so that Plato and Aristotle, and even Plotinus living in the Christian era, were seen as having their place in God's economy. There is only one overriding exclusivism – the exclusive demands of the search for truth. Within that search, I need to hold fast to what an Indian theologian, Pandipeddi Chenchiah, called 'the raw fact of Christ'.[17]

Vernon White encourages us to think of God's revelation in terms of divine saving action rather than in terms of propositional knowledge. He says 'epistemology must not hijack ontology'.[18] '*Knowledge* of the Saviour is not a necessary constituent of *being* saved: not, that is, in this life, and not in

---

[14] Quoted in D'Costa (1986), p. 59.
[15] Moltmann (1977), p. 153.
[16] D'Costa (1986), p. 80.
[17] Quoted in Cragg (1986), p. 208.
[18] White (1991), p. 24.

the sense that historical knowledge about the events of Jesus of Nazareth is required.'[19] He does not deny the decisive significance of the history of that life and death and resurrection (indeed it is the whole point of the book from which these quotations are drawn to seek to understand the atonement and the incarnation), but nor does he deny a degree of cultural conditioning of the transmission to us of our knowledge of the unique Christ-event:

> The historical event of Jesus Christ can remain pivotal, without for one moment claiming epistemological purity or exclusiveness. It may be a unique focus of religious truth without for one moment denying that light is also scattered further afield. Such light will not only demand recognition in its own right, it may well illuminate and correct the distortions which collect even around its focus.[20]

White's account is centred on God's saving action, and one must distinguish this from those pluralist accounts which are centred on the possibilities of human self-fulfilment through varieties of salvific practice. Just as I reject mere instrumentalism in science (since science's ability to get things done must surely depend upon some verisimilitudinous description of how things are), so I do not think that an account of religion merely in terms of culturally differing techniques of fulfilment can be adequate, since religious *praxis* must bear some consonant relationship with the Reality to which religion testifies.

An inclusivist position confers no monopoly of truth on any tradition, and therefore it encourages dialogue between the traditions. Pannenberg says briskly, 'Only the positivity of a commitment in faith which isolates itself from critical thought can simply identify the revelation of God in its own tradition as opposed to others . . . There is no surer way of abandoning the priority of the reality of God over the arbitrariness of pious subjectivity.'[21] The increasing pressure of secularization faced by all religions can only make the task of encounter more pressing. Cragg says that 'The faiths have never had more urgent tasks with themselves than those which the world now sets each of them alike. Theology, therefore, never had more reason to be ecumenical.'[22] His book is a sustained exercise of theology in 'cross-reference'. Yet dialogue must be prepared to accept the vulnerability required for its reality, while its participants hold fast to their cherished insights in what Buber called 'speech from certainty to certainty', in a situation of 'anguish and expectation'.[23] It will be a painful exercise. John

[19] ibid., p. 112.
[20] ibid.
[21] Pannenberg (1976), pp. 319–20.
[22] Cragg (1986), p. 345.
[23] Quoted in D'Costa (1990), p. 190.

Milbank goes so far as to say, 'It will be better to replace "dialogue" with "mutual suspicion",'[24] but I think that is too harsh a judgement. The encounter must involve eschewing all easy caricature and polemic. 'A wise inter-theology will know that all faiths can be controversially embarrassed if we are crude enough to misconstrue them.'[25] It is only too easy to turn Christian trinitarianism into tritheism, Hindu practice into a merely animistic polytheism.

Each tradition may expect enhancement and correction from the others. So staunch a defender of Christian uniqueness as Lesslie Newbigin is content to say, 'The Holy Spirit, who convicts the world of sin, of righteousness and of judgement, may use the non-Christian partner to convict the Church.'[26] In fact, the Christian understanding of the hidden work of the Spirit (chapter 8) may be the best way, within a trinitarian understanding, of giving an account of God's presence and activity throughout the world faiths.[27]

Although Keith Ward, in an interesting, if selective, survey of seminal writings from four religious traditions, concludes that he detects signs of a common 'iconic vision', so that 'To see the temporal as the image of eternity and to pursue the path of self-transcendence in relation to it is perhaps the central clue to understanding the religious form of life,'[28] yet he also acknowledges that 'Religions generate infinite differences. They often come into existence as a reaction against other existing religions.'[29] In the opinion of John V. Taylor, 'The deeper the appreciation of the other faith the greater the knowledge of what these disagreements really are.'[30] Pannenberg's thinking assigns great significance to the history of the religions, but he does not disguise their differences: 'Fidelity to the nature of the world religious phenomenon itself . . . demands that the conflict of the Christian truth claim with those other religions be made explicit. Yet even Christianity falls under the provisionality of the present.'[31] It is time to consider such points of conflict in greater detail. I do so conscious of my ignorance, and also of how hard it would be, even after much study, properly to understand a religious tradition that had not been the foundation of one's own way of life. Nevertheless I must make the best attempt I can to give some consideration to these problems.

Let us begin with the individual human self, seen by all the religions of the

[24] Quoted in D'Costa (1986), p. 121.
[25] Cragg (1986), p. 18.
[26] Quoted in D'Costa (1986), p. 136.
[27] See Barnes (1989), especially ch. 7.
[28] Ward (1987), p. 156.
[29] ibid., p. 1.
[30] Taylor (1972), p. 186.
[31] Grenz (1990), p. 35.

Near East (Judaism, Christianity, Islam) as being of infinite value in the sight of the God of Abraham, Isaac and Jacob, and seen by Buddhism, with its doctrine of *anatta* (non-self), as an illusory individualism which produces *dukkha* (suffering). Of course, Christianity recognizes the ambiguous character of the ego apart from God. We recall the stern words of Jesus about denying oneself and taking up the cross (Mark 8.34–6, par.), and the words of Paul that 'it is no longer I who live, but Christ who lives in me' (Gal. 2.20), but their force is to encourage the hope of a true fulfilment in transcended selfishness, not in self-extinction – to gain one's life through losing it to Christ. The Christian diagnosis of the human condition sees sin (alienation from God) as the root of the problem, and a repentant and faithful return to God as its solution. The Christian hope lies not in the attainment of non-desire, but in a purification that leads to right desire, that seeking of the soul for God which is central to the thought of Augustine, that dart of longing love by which the Christian mystic seeks to penetrate the Cloud of Unknowing. Cragg is surely right to say, 'It is clear that Christian faith has a radical quarrel with the Buddhist equation between selfhood and selfishness.'[32] In any dialogue I must take my stand humbly but firmly on the side of the worth of human individuality. David Jenkins says, sternly but truly, of the Eastern ideal of escape from the self, that it 'is clearly an acknowledgement of the total bankruptcy of man. The point of being human is to cease to be human.'[33] The Christian tradition speaks rather of 'the glory of man'.

The Near Eastern religions concur in assigning a linear, progressive significance to time, within which God's purposes are in course of their outworking. To the Far Eastern religions, it seems, time is but the circling of samsaric flux. Is time a path to be trodden or a wheel from which to seek release? I cannot but take the former view. It leads to a strongly positive evaluation of history, hence my concern (chapter 5) with the historical Jesus. By contrast, Gandhi wrote:

> I have never been interested in a historical Jesus, I should not care if it was proved by someone that the man called Jesus never lived, and what was narrated in the Gospels was a figment of the writer's [*sic*] imagination. For the Sermon on the Mount would still be true to me.[34]

Of course, I am not chaining myself to a prosy literalism and denying the power of poetic myth. Ward is doubtless right to say that 'it is a major heresy of post-Enlightenment rationalism to try to turn poetry into pseudo-

[32] Cragg (1986), p. 255.
[33] Jenkins (1967), pp. 72–3.
[34] Quoted in Robinson (1979), p. 50.

science, to turn the images of religion, whose function is to evoke eternity, into mundane descriptions of improbable facts'.[35] But it is the realized myth of the incarnation, the Eternal made known in a birth under Augustus and a death under Tiberius, which stands at the heart of my own belief. Hans Küng is right to say that 'The Christ-event is bound up with *concrete history* in a way quite different from the Buddha-event.'[36] Christianity is inescapably concerned with history. Its stories are not merely pragmatically useful encouragements to right living.

All traditions recognize the difficulty of talk about God (or whatever form ultimate reality is held to take). There is an inescapable tension between apophatic theology (acknowledging the mysterious ineffability of God) and cataphatic theology (which nevertheless seeks to say something about that mystery since, as Augustine said, we cannot be content merely to be silent). Aquinas asserted, 'We cannot know what God is, but only what he is not.'[37] Yet there is a deep religious longing to know what God is like, to answer the question, Is reality ultimately gracious to us? We have to traverse the theological razor's edge between talkativeness and taciturnity. Of course, the essence of the Infinite is beyond the reach of our finite minds. It is doubtless true that 'It is important to grasp the extent to which agnosticism lies at the heart of orthodoxy.'[38] It may well be the case that a great deal of theological discussion is aimed more at fencing off error than at adequate articulation of the truth. Yet no religious tradition is without its body of teaching and practice, which constitute its insight into the way things ultimately are.

All traditions possess scripture: writings held to be authoritative and illuminating. Part of the ecumenical encounter of the world faiths must involve decisions about how they are to view each other's sacred writings. My inclusivist stance implies for me a clear primacy for the New Testament. The unique role of Christ gives that body of writings a correspondingly unique importance. The Old Testament (including all its primitive passages and unedifying stories) is scripture for me because it was scripture for Jesus and the first disciples. My inclusivist stance equally implies that I may expect to find passages of spiritual value, including, doubtless, insights not to be found elsewhere, in the other scriptures of the world. Cragg speaks of 'the severe geographical limitations of the New Testament',[39] and he raises the question of whether the canon was closed prematurely. Christianity had encountered the thought of Greece but not yet that of India. If one believes

[35] Ward (1987), p. 3.
[36] Küng (1986), p. 432.
[37] Quoted in Ward (1987), p. 135.
[38] ibid., p. 115.
[39] Cragg (1986), p. 328.

in God's providential hand in history, the form of the encounters of early Christianity, and the Hebraic matrix which gave it birth, will not be just accidental happenings. To say that is not to deny the possibility of further enlightenment, but it will surely be in the mode of the exploration of the foundational New Testament witness from new perspectives. Bishop Westcott said that the greatest commentary on St John would be written by an Indian.

But what are we to make of other scriptures, for example, the *Bhagavad Gita* and the Qur'an? Formidable linguistic and hermeneutical barriers have to be surmounted if any genuine contact is to be made. I wish to be respectful, but I have to say personally that they do not speak to me as the Bible does. I am sure that these problems will not be solved by eclectic appropriation of congenial passages plucked from the scriptures of other faiths. I have to confess to considerable reservations about occasions of inter-faith worship in which common denominator lessons are read. I think the faiths have to meet each other in the strangeness of their differences, and if any common form of worship is attempted it might well be best conducted in a shared meditative silence.[40]

That reminds us that there are present in all religious traditions claims of mystical encounter, direct experiences of union with the All or with the One. I understand them to be true encounters with God in his immanence, either interpreted personally (the One) or impersonally (the All). The debasement of the word 'mysticism' to mean the vaguely religious or cloudily questionable is greatly to be deplored. We need a specific term for this well-documented phenomenon. Although mysticism is more frequently associated with the religions of the Far East, it is also to be found in the Near Eastern religions. Despite Küng's quoting with approval Hegel's remark that 'there is no mystical "night in which all cows are black" ',[41] it does seem to me, to the small extent to which I understand the matter, that the unitive experience of the mystics is one in which the traditions draw close to each other. Küng is cool about the ladder of mystical prayer, saying, 'if we take the New Testament as our critical standard, such a scheme of prayer will never have any sort of definitive validity for Christians'.[42] I am not sure that St Paul (2 Cor. 12.3–4) would agree with him.

We must not over-emphasize the conflict of traditions. The boundaries between them are fractal-like (infinitely interpenetrating). A facile charac-terization of Buddhism as world-denying, Christianity as world-affirming, comes up against the facts of Bodhisattvas and the Desert Fathers. Simple

[40] cf. Cracknell (1986), pp. 133–8.
[41] Küng (1986), p. 174.
[42] ibid., p. 423.

religious type-casting fails to do justice to the subtleties involved. The Christian Meister Eckhart seemed to speak in monist terms (identifying God and the deep self); the Hindu Gandhi was deeply involved in political action. There are common motifs which recur, such as that of path or way.[43]

Yet there is that great point of difference between us when the traditions come to consider their attitude to Jesus. Many outside Christianity are happy to accord him a degree of recognition and to exhibit a sympathetic appreciation of his teaching, but they would not be able to speak of him, as I have done in these lectures, as being the unique incarnation of God in humanity. The Qur'an grants Jesus the status of a prophet, but it also says, 'People of the Book, go not beyond the bounds in your religion . . . The Messiah, Jesus son of Mary, was only the Messenger of God and of His Word that He committed to Mary, and Spirit from Him. So believe in God and His Messenger and say not "Three" ' (*Sura* 4.169). Muslims believe in Jesus' virginal conception, but they are also taught by the Qur'an that he did not die on the cross. Such an end was unfitting for one of God's Messengers, so Jesus was spirited away to heaven and someone else died in his place. Not only is this notion totally unhistorical, but it also is in contradiction with that profound and fundamental insight of Christianity that the cross of Christ represents God's solidarity with us as a 'fellow sufferer'. This literally crucial issue is one that any realistic inter-faith dialogue will have to face, for the cross is truly a *skandalon*, a 'stumbling block' (1 Cor. 1.23). Aelred Graham wrote of Buddhist reactions to the crucifix: 'The depiction of an almost naked human being nailed to a cross as an object to be revered is unintelligible, with its presentation of barbarous cruelty, to a devout Buddhist, who is disposed to regard violent death in any form as the result of "bad karma".'[44] Even in Amida or Pure Land Buddhism, often thought to be the strand of that tradition closest to Christianity, the concept is of salvation through shared merit and not vicarious suffering. The typical Buddhist symbol is that of the serene and smiling Buddha, seated in the lotus position – not a man on the gallows.

One could think that Hindu thought might prove more hospitable. Does it not have the concept of *Avataras*, incarnations of Lord Krishna taking place at points of time where humanity stands in special need of divine assistance? Yet not only are these incursions repeated (while the life of Christ is once for all), they are also more of the nature of docetic appearances than true engagements with human history. Robinson writes of an avatar:

> he enters the process and passes from it without creating a ripple. 'There is an old Hindu tradition that when an *avatara* walks his feet do not touch

[43] See Cracknell (1986), pp. 75–85.
[44] Quoted in Robinson (1979), p. 93.

185

the ground so that he leaves no footprints.' Even though he may die, as Krishna does, his death does not affect or effect anything. He is not one of us . . . but Vishnu making an appearance in the form of a man.[45]

An Indian Christian, Vengal Chakkarai, has sought to locate the cross in Indian thought, saying that in its darkness and dereliction Jesus is completely emptied of self, 'plunged into *nirvana* or *suniam* where God is not', and so becomes 'the most ego-less person known in history, and therefore the most universal of all'.[46] There is something here for us to ponder, but I see great difficulty in reconciling this insight with the resurrection, the exaltation and the glorification of the *individual* Jesus, bearing the scars of his passion.

I feel I have to say that, from a Christian perspective, tales of divine appearances which we find in the legends and stories of other traditions appear as hints and premonitions of God's great historic act in Christ. That may sound patronizing, but loyalty to the truth as I perceive it gives me no option. The uniqueness of that revelatory act (for, 'To suppose that God might have several human faces is to lose the real revelatory significance of the Incarnation'[47]) imposes an inescapable scandal of particularity. I cannot agree with Maurice Wiles that 'An understanding of incarnation which emphasizes its discontinuity and exclusivity in relation to the wider religious history of the human race, can have no place in and leaves no place for a theology of dialogue.'[48] The only basis for honest dialogue is the humble but firm presentation of one's own understanding of truth. A consensus purchased by down-playing the status of Jesus, which these lectures have sought to defend, would not be worth having. For me, Jesus' resurrection (usually neglected by those seeking a pan-religious synthesis) is sufficient in itself to mark him out as historically unique.

Although Judaism and Islam evaluate Jesus so differently from Christianity, it is clear that the Near Eastern religions are in other respects quite close to each other, as their historical associations would lead us to expect. They see God as One, the Creator of the world, active in its history, merciful in his dealings with those who seek him. That is a substantial common basis on which to found a dialogue. Yet other differences remain.

If Jesus' status as God's Son is non-negotiable for Christianity, Israel's status as God's people is non-negotiable for Judaism. Jacob Neusner wrote:

For my part, if I am accepted as a 'human being' and not as a Jew, I do not accept that acceptance. I aspire to no place in an undifferentiated

[45] ibid., p. 48.
[46] Quoted in ibid., p. 123.
[47] Hebblethwaite (1987), p. 7.
[48] Wiles (1992), p. 70.

humanity. Take me despite my Jewishness and there is nothing to take.

To which Kenneth Cragg proposes the Christian response:

> If I am only accepted as a 'Gentile' I do not accept that acceptance. I aspire to no place in a differentiated humanity if the differentiation is radical and not merely those creaturely things which are natural, cultural and incidental. Take me despite my 'Gentileness' and there is everything to share.[49]

No Christian should deny God's special relationship with his ancient people, who were the vehicle of his historic revelation and the community from which Jesus sprang. Yet those differences between Jew and Gentile we understand now to be transcended in the new fact of Christ (Gal. 3.28; Eph. 2.14–15). What that means in relation to all that valid experience of God preserved in Judaism is a question with which Paul wrestles in Romans 9–11. Rather than the currently fashionable proposition that Jews have no need to learn anything from Christians, I prefer his solution:

> I ask, then, has God rejected his people? By no means! . . . Now if their trespass means riches for the world, and if their failure means riches for the Gentiles, how much more will their full inclusion mean! . . . For if their rejection means the reconciliation of the world, what will their acceptance mean but life from the dead? . . . O the depth of the riches and wisdom and knowledge of God! How unsearchable are his judgements and how inscrutable his ways! (Rom. 11.1, 12, 15, 33)

I believe the former view arises from a deep feeling of guilt for the many Christian centuries of anti-Semitism and for the consequent Christian share in responsibility for the horror of the Holocaust (*Shoah*). No Christian can fail to express abhorrence of anti-Semitism and all its evils, and a profound penitence for the dreadful deeds of the past. But however rightly and strongly we feel that, it does not mean that it should negate our witness to what we believe to be the truth. Humbly, but persistently, we have to present that truth as we see it to our Jewish brothers and sisters.

Christianity shares with Judaism (from where it learned it) the hope of God's Messiah, his anointed one who is the instrument of his saving purpose. The expectation of 'He that was to come' is a point of meeting between the two traditions.[50] Christianity believes that expectation to have been fulfilled in Jesus Christ, and it must, in honesty, testify to that belief. Judaism cannot

---

[49] Cragg (1986), pp. 140–1.
[50] ibid., p. 97.

accept that this is so.[51] Indeed, it has been suggested that to be *awaiting* the Messiah is an essential part of Judaism.[52] In rejecting the Messiahship of Jesus, Judaism puts a sharp question to Christianity: If the one who was to come has actually come, how is it that his coming has made so little difference to the world? History AD seems as full of unmerited suffering as ever it was BC. The bitter sadness of the *Shoah* can only intensify that feeling. Buber says:

> I believe the world's redemption did *not* become a fact nineteen centuries ago. We are still living in a world that is not redeemed . . . The man who raises Jesus to so high a place ceases to be one of us, and if he wants to challenge our faith that redemption lies in the future then we go our separate ways.[53]

No Christian could deny that 'we do not yet see everything in subjection' to Christ, that God's kingdom is not now manifested on earth. However, in the Christian paradox of 'already but not yet', we do believe we see the resurrection of Christ as the victory from which all hope of redemption springs, and so we look to Jesus, 'crowned with glory and honour because of the suffering of death, so that by the grace of God he might taste death for every one' (Heb. 2.8–9).

The non-negotiable hard core of Islam is its belief in the supreme authority of the Qur'an. Muslims are People of the Book in a much stronger sense than Jews or Christians. In their view, the Qur'an is not a divinely inspired set of human writings, but a divinely dictated document. Islam is a verbal religion, centred on a literal conception of the word of God. The prophets are messengers, divine mouthpieces. What they proclaim is what matters. The recognition in Israel, which we see with Jeremiah and Hosea, that the life of the prophet is also part of what God is doing through him, is foreign to Islam. Hence the recoil from notions of incarnation, of the Word made flesh. That rejection of incarnation is confirmed by Islam's intense feeling for that utter transcendence of Allah. Although Allah is the all-merciful, the Islamic keynote is power, not grace. Its optimism is focused in the theocratic state, the way of life conducted in accordance with the *Shari'ah*, the Islamic law.

This poses many problems for the Christian, but not as great as those encountered when we turn from the religions of the Near East to those of the Far East. Much of their world of thought is difficult for an occidental to penetrate, and, in my ignorance, I must restrict myself to some simple observations.

[51] See Lapide (1984).
[52] Cragg (1986), p. 15.
[53] Quoted in ibid., p. 114.

Hinduism is a Western term for the venerable, sprawling, diversified traditions of indigenous Indian religion. Its fertility and variety are such as to defy generalizations about it. 'Understanding Hinduism might be likened to embracing a cloud. It escapes you as you will to grasp it.'[54] It seems the most syncretistic of the world faiths, welcoming into its capacious bosom all the insights that come to it. 'There is a long Hindu tradition of absorption, of non-distinction, of reverencing reverence *per se*, of tolerating all but intolerance.'[55] Hinduism is quite prepared to take Christ on board, but not his uniqueness: 'What India wants to challenge is monopoly.'[56] Of course, I find it impossible to consider Jesus as one avatar among many.

Great difficulty is presented by the doctrine of *samsara*, the continuing flux of reincarnation. This seems a natural notion to the Indian mind, not worth serious questioning. To the Western mind, imbued with the idea of the individual formed by a unique heredity and experience, it does not even appear a coherent concept. What 'spiritual ingredient' is it which provides the continuity between an Egyptian slave and me? I suppose part of the Hindu answer would be *karma*, the entail of good and evil from one life to the next. It is clear that one of the attractions of reincarnation lies in its furnishing a ready-made theodicy: present suffering is the consequence of past misdeeds, in an unverifiable way. However, Küng comments that 'the question of theodicy is not really answered, just postponed';[57] it is carried over from one generation to the next. Reincarnation offers a tactical solution to the problem of how individual suffering is distributed, but not a strategic solution to the problem of why there is a world so constituted as to have so much suffering in it. Gorringe refers to the missionary A. G. Hogg as concluding that 'if the Law of Karma is true, then history is robbed of its deepest meaning'.[58] It becomes the outworking of a remorseless destiny, the spinning of a wheel of fate. Redemption is then held to lie in attaining release from the samsaric cycle. Ward comments, 'All this presupposes that rebirth is a thing which is in the end undesirable; and individuality possesses no unique value, worth preserving.'[59] It is a far cry from the God of Abraham, Isaac and Jacob.

Hindu thought is strongly monist. It not only sees the divisions of individual selves as fundamentally illusory (an illusion, it seems, which is recycled by *samsara*), but an equal illusion is the division between God and his

[54] ibid., p. 17.
[55] ibid., p. 177.
[56] ibid., p. 16.
[57] Küng (1986), p. 232.
[58] Gorringe (1991), p. 11.
[59] Ward (1987), p. 20.

world. The Ultimate is seen as *nirguna Brahman*, utterly without differentiation or attributes, as in the stern *advaita* (pure non-dualism) of Sankara, though this was modified in the subsequent thought of Ramanuja. However, Hindu practice is not so austere, and looks to *saguna Brahman*, reality with qualities. Ward defines an 'operative similarity', based on how concepts actually function in the life of a believer, and he concludes his survey of these perplexing matters by claiming that 'the Vedantic doctrine of Brahman is operatively similar to the orthodox doctrines of God in Middle Eastern traditions'.[60] That would be a very important, if surprising, conclusion were it to be sustained by further research.

Even further from Western religious thought is the world-view of Buddhism. Cragg says that 'Perhaps more than any other faith Buddhism requires to be met only on its own ground.'[61] In its eyes, its questioners have not yet renounced the illusions in which they are trapped. At its centre is the doctrine of *anatta* (non-self). 'All things are impermanent, made of collections of individual fleeting events. There is no enduring substantial self, underlying its fleeting thoughts and feelings and perceptions.'[62] (There are strange resonances here with the event-dominated process metaphysics of Whitehead, pushed to its limits.) If there is no self, in Theravada Buddhism there is also no God. 'The central concept of Buddhism is not God but nirvana,'[63] itself a concept elusive and indefinable. Küng's verdict is that 'Buddhism, if not atheistic, is nonetheless decidedly *agnostic*';[64] Gautama refused to speculate on metaphysical questions of God or the origin of the world. Perhaps that is why Buddhist thought, at least in a Western transposition, has proved surprisingly attractive to European and American intellectuals. It is also a practical religion, concerned with the eightfold way by which life should be lived. In its home countries it is strongly monastic. It is in the community of the *sangha* that enlightenment is to be sought. There is no elitist taint in this monasticism (as there was in medieval Christianity), because with the turning of the samsaric wheel everyone's turn can come.

This all too simple survey of the great traditions, however inadequate, makes clear something of the huge task involved in the ecumenical encounter of the world faiths. It will be a work of centuries. I hope that Christianity will participate in that dialogue in a way that is respectful but not too apologetic or appeasing. Integrity requires us to speak for the truth as we see it, and we need to remember Farrer's warning that 'the

[60] ibid., p. 50.
[61] Cragg (1986), p. 249.
[62] Ward (1987), pp. 61–2.
[63] ibid., p. 51.
[64] Küng (1986), p. 390.

acknowledgement of a vital truth is always divisive until it becomes universal'.[65] I sometimes fear that Christianity is a little too eager for dialogue, a little lacking in nerve to hold fast to what it has learned of God in Christ. We Europeans must shake off lingering guilt arising from our colonial past. We certainly do not want to be triumphalist, but nor do we wish to forget that there may well be issues on which we are right and those who do not share our view are mistaken. In the end, it is the question of truth that matters, and there is an inevitable exclusivity about truth. If you tell me that you hold the view that the phenomenon of heat is due to the subtle fluid caloric, I do not say that you are entitled to your opinion and I respect you for it. I try, instead, to convince you of the correctness of the kinetic theory of heat energy. Either Jesus is God's Lord and Christ or he is not, and it matters supremely to know which is the right judgement. Of course, we must be careful to distinguish between the necessary intolerance that truth has of error and a social intolerance exhibited by failing to respect as people those whose opinions we believe to be mistaken. I do not despise you for your caloric belief, nor do I try to impose my kinetic belief on you by harassment or manipulation. This important distinction is one that many seem to find difficult to make when considering relationships between different faith-communities. We *can* conduct our dialogue both with respect and to the point. It has been suggested that Christianity has, by its speaking of both the normative particularity of Jesus and the universal, diverse working of the Spirit, a particular contribution to make to the possibility of true encounter between the faiths. D'Costa says, 'I believe that the Trinitarian doctrine of God facilitates an authentically Christian response to the world religions because it takes the particularities of history entirely seriously.'[66] A religion which has resisted its own dissolving into a gnostic account of timeless truth should be open to meeting the historic idiosyncrasies present in all religious traditions, without reducing them to merely contingent collections of opinion.

In relation to the question of truth, I am conscious also that there are apparent conflicts between the scientific world-view and the kind of metaphysics it might be thought to support, and the world-views of the Far East. Modern science came into being in the sternly realist setting of Near Eastern religious thought, whose shared concepts of God encouraged the view that the physical world should reflect in its order the rationality of its Creator, while his freedom in creating implied that one must interrogate that world to find out what actual form its contingent order had taken.[67]

[65] Quoted in Hebblethwaite (1987), p. 123.
[66] D'Costa (1990), p. 17.
[67] Jaki (1978); Russell (1986).

I am aware that there are those like Fritjof Capra who hold the opinion that the subatomic world of modern physics is one that is startlingly in accord with the descriptions of Eastern thought.[68] I have previously given my reasons for dissenting.[69] The quantum world is subtle and elusive, but it is not wholly dissolving. There are also persistent patterns, which we describe through symmetries, and there are conservation laws which control and preserve within the flux of process. The mutuality of bootstrap dynamics (everything is made of everything), which Capra espoused, has not proved to be the case. There are identifiable constituents (the quarks and gluons). The quantum world is certainly interconnected, with its counterintuitive togetherness-in-separation,[70] but relativity theory also prescribes a degree of causal independence for spatially separated systems. As observers we exert some influence on the world that we experience (to a degree which is still a matter of contention), but that does not give us licence to regard it as a world of appearances, the result of the play of *maya* (illusion).

There are very difficult unsolved problems in interpreting quantum theory and assessing its metaphysical significance.[71] I do not think that Eastern thought solves these problems for us. I also suspect that where it has been claimed to do so, it is in fact a Westernized, even Californian, selective version of that thought which is proffered to us. Küng says that Capra 'has to concentrate, in a vague, eclectic manner, on certain selected philosophical aspects of Eastern mystical thought without always paying sufficient heed to their complexity and even contradictoriness'.[72]

I do believe, however, that these questions of physical reality, and the related questions of natural theology which arise in an unforced way from them, constitute a modest but promising area of interaction for the world faiths. Issues here are capable of easier clarification, and are less fraught with threats than those which lie closer to the heart of what is vital to each tradition. I want to ask my colleagues in Eastern religions what they make of the finely tuned balance in the laws of nature, enabling the evolving history of the universe to achieve its astonishing fruitfulness, which the scientific insights collected under the rubric of the Anthropic Principle present for our consideration. In this way we humble bottom-up thinkers might make our modest contribution even to the ecumenical encounter of the world's great religious traditions.[73]

---

[68] Capra (1975); Zukav (1979).
[69] Polkinghorne (1986), pp. 82–3; (1988), pp. 93–4.
[70] Polkinghorne (1984), ch. 7.
[71] d'Espagnat (1989); Polkinghorne (1991), ch. 7.
[72] Küng (1986), p. 379.
[73] cf. Rolston (1987), pp. 258–69.

Epilogue

Scientists who are hostile to religion tend to make remarks such as 'Unlike science, religion is based on unquestioning certainties.'[1] They thereby betray their lack of acquaintance with the practice of religion. Periods of doubt and perplexity have a well-documented role in spiritual development. 'Faith seeking understanding' (in Anselm's splendid phrase) is the quest for motivated belief. Like the scientific quest, that journey of intellectual discovery is made by those who survey experience from an initially chosen point of view, which must be open to correction in the light of further experience. Religion has long known that ultimately every human image of God proves to be an inadequate idol.

Lewis Wolpert asks the question, 'Why should religious experience be treated as different from any other experience and not subject to scientific inquiry in the normal way?'[2] The answer is that all experience is to be subject to rational inquiry, and part of that necessary rationality is to conform one's investigation to the nature of the entity being investigated. I very much doubt whether Professor Wolpert subjects his enjoyment of music or his encounter with persons 'to scientific inquiry in the normal way', if that phrase is to be interpreted in some flat, universal catch-all, reductionist way.

Religion does not demand that all answers are agreed before the discussion begins. All that it asks for is a respect for its particular modes of experience and an openness to the insights they bestow. This series of Gifford Lectures, like so many of its predecessors, has been an exercise in just such a search for the grounds of motivated religious belief. I see no fundamental incompatibility between any of my bottom-up searchings for the truth, whether pursued in the field of science or in the field of Christian belief. Both domains of inquiry are necessary if we are truly to comprehend the way things are. Many puzzles remain, but the scientist and the theologian can make common cause in the search for understand-

---

[1] Wolpert (1992), p. 144.
[2] ibid., p. 147.

ing, pursued with openness, scrupulosity and humility, conscious (in the words of Isaac Newton, for whom both science and religion were matters of supreme concern) of the great ocean of truth lying undiscovered before us.

Glossary

In the text, reference is made to ideas, both scientific and theological, which may not be familiar to all readers. It would interrupt the flow of argument to explain each as it occurs, so the following brief notes are offered as modest aids to comprehension.

adoptionism The theory that divine powers and status were conferred on the man Jesus, either at some stage of his life or at his resurrection, so that he became the adopted Son of God.

anthropic principle A collection of scientific insights indicating that the possibility of the evolution of carbon-based life depended upon a very delicate balance among the basic forces of nature and (perhaps) also on very specific initial circumstances for the universe.

Apollinarianism The theory that in Christ the human mind was replaced by the divine Logos.

Aramaic The Semitic language, closely allied to Hebrew, which was spoken in Palestine in the first century.

Arianism The theory that Christ was a being of semi-divine dignity, created by God the Father before the world began as a kind of half-way house between divinity and humanity.

aseity The quality of existing from itself alone possessed by a being who needs no other explanation.

asymptotic The property of behaviour emerging in some extreme limiting situation.

background radiation The cold radio-noise which fills the universe as a relic of the state of affairs when radiation ceased to interact with matter at a cosmic age of about half a million years after the big bang. It is a kind of cosmic fossil.

Barth, Karl A dominant figure of twentieth-century theology, his 'crisis theology' was a reaction against liberalism and asserted that God is known solely through the disclosure of his Word.

big bang That singular event of infinite density from which the observable universe appears to have originated approximately fifteen billion years ago.

bootstrap In modern physics a particle can play two roles. It can be the result of a combining together of other particles, and also its exchange between those other particles can create the force that holds them together to make itself. In other words, it can be the cause of its own effect. This led to the bold conjecture that there are no basic constituents, but 'everything is made of everything' in a self-sustaining act of self-consistency. This pulling of the world into existence by its own bootstraps has not proved to be a successful physical theory, but it remains of conceptual interest.

chance and necessity The pattern of evolution appears to be that happenstance (e.g., a genetic mutation) is the origin of novelty, which is then sifted and preserved in a lawfully regular environment (e.g., the reasonably reliable transmission of genes from one generation to another, together with the operation of natural selection in a relatively stable environment). This fruitful interplay is summarized in the slogan 'chance and necessity'. Both aspects are essential to its operation.

chaos theory Physical systems which are non-linear (doubling the cause produces something more complicated than just twice the effect) and reflexive (the system can act back upon itself), often display such an exquisite sensitivity to their detailed circumstance that their behaviour becomes intrinsically unpredictable. The way they behave is not unrestrictedly haphazard, however, but they explore a confined range of possibilities (called a 'strange attractor'). Thus chaotic systems exhibit a kind of ordered-disorder. The recognition of this behaviour is comparatively recent, and the study of chaos theory is still at an early stage of development. Some regard it as a 'third revolution' in physics, comparable to the discoveries of Newtonian mechanics and quantum theory.

Christology The theological discipline concerned with understanding the nature of Christ, and in particular the way and degree to which both human and divine characteristics are present in him.

complementarity An insight of quantum theory (particularly associated with the thought of Niels Bohr) that there are alternative and mutually exclusive modes of description of the same entity (cf. wave and particle, below).

Ebionitism An adoptionist (*q.v.*) account of Christ, stressing his humanity, associated with certain early Jewish Christians.

economy A term used by the Greek Fathers of the Eastern Church to indicate divine interaction with the created order (as opposed to the divine nature in itself).

entropy The measure of the disorder of a system. The second law of thermodynamics asserts that in an isolated system (insulated from external influence) entropy can never decrease (things tend to become more disorderly).

epistemology That branch of philosophy concerned with the theory of knowledge and how it is attained.

EPR experiment A counterintuitive 'togetherness-in-separation', by which quantum entities which have once interacted with each other retain a power of instantaneous mutual influence, however far apart they separate. It was first identified by Einstein, Podolsky and Rosen, and it has been experimentally confirmed by Aspect and his collaborators.

eschatology The theological discipline concerned with 'the last things', that is, how history will end and what is creation's ultimate destiny.

196

form criticism The study of the nature of the literary elements (pericopes) making up a text (particularly the gospels) with regard to their character (healing stories, sayings, etc.), origin and social setting, transmission and development (particularly in the oral tradition preceding the written form).

gnosticism Religious thought that lays particular emphasis on knowledge and enlightenment. There were such movements in the early Christian centuries with a particular concern to gain thereby release from the flesh and from the influence of cosmic powers. Their origin is disputed, but they were regarded as heretical by catholic Christianity.

God of the gaps The invocation of God as an explanation of last resort to deal with questions of current (often scientific) ignorance. ('Only God can bring life out of inanimate matter,' etc.). Such a notion of deity is theologically inadequate, not least because it is subject to continual decay with the advance of knowledge.

Gödel's theorem The logical discovery that systems of sufficient complexity to contain the whole numbers are always incomplete, in the sense that they contain propositions which are stateable but not provable within that system.

hermeneutics The theory of the interpretation of texts. Central to its concern is the question of how the original meaning and a contemporary understanding of the text should relate to each other.

kenosis Self-emptying, particularly the self-limitation accepted by God in his acts of creation and incarnation.

Laplace, Pierre Simon The greatest of Newton's successors. He was deeply impressed with the apparent determinism of the Newtonian system, so that he wrote: 'If we conceive of an Intelligence who, for a given moment, embraces all the relations of being in the Universe, it will also be able to determine for any instant of the past or future their respective positions, motions and generally all their affections.'

many-worlds quantum theory The interpretation of quantum theory that supposes that all the possible outcomes of a measurement (*q.v.*) in fact occur in parallel disconnected worlds into which physical reality divides at each such act of measurement.

measurement problem Quantum theory in general only predicts the relative probabilities of a variety of different outcomes of each act of measurement. It is an unresolved interpretative problem how it comes about that an observer records a particular such outcome in any single act of measurement.

Michelson–Morley experiment Attempts from 1881 onwards by the two physicists A. A. Michelson and W. W. Morley to measure the Earth's velocity through the supposed aether. All attempts gave a null result. The accepted explanation is given by Einstein's theory of special relativity.

miracle Etymologically, a miracle is a highly surprising event (which could include the exercise of unusual human powers or a significant coincidence), but it is commonly used in the sense due to David Hume of 'the violation of a law of nature by a god'. In the view of the present author, a more carefully nuanced account is necessary (see pp. 103–4).

natural theology The attempt to learn something of God by the exercise of reason or from the consideration of general experience. In these lectures the meaning is extended to include an appeal to particular experience, treated as evidence to be interrogated and not as unquestionable revelation.

Nestorianism An account of Christ which treats him as the concurrence of two persons, one human and one divine.

non-locality see: EPR experiment.

ontology That branch of philosophy concerned with being, and so with what is the case.

panentheism The theological concept that the world is part of God but God exceeds the world (cf. pantheism, which equates the world and God).

pneumatology The theological discipline concerned with the nature and activity of the Holy Spirit.

process philosophy The philosophical system developed by A. N. Whitehead as a metaphysics based upon events as fundamental entities. Charles Hartshorne and others developed process theology upon this philosophical foundation.

Q A hypothetical document held by some New Testament scholars to be the basis of material, much of it of close verbal similarity, common to the gospels of Matthew and Luke and not found in Mark.

quantum cosmology Conjectural theories of the very early universe in which the effects of both gravity and quantum mechanics would have been simultaneously significant. The current absence of a consistent theory of quantum gravity makes such imaginative attempts necessarily speculative.

quarks and gluons The currently accepted basic constituents of nuclear matter.

redaction criticism The discipline of New Testament study which concentrates on understanding each gospel from the presumed point of view of its author and the community in which the material was edited.

sociobiology An account of behaviour (particularly human behaviour) seen as arising principally from genetic influences.

soteriology The theological discipline concerned with understanding salvation, conceived in a variety of ways such as deliverance from evil, restoration of integrity, and ultimate fulfilment.

spontaneous symmetry breaking An important scientific insight that the solutions of a theory can be less highly symmetrical than the theory itself; e.g., a pencil balanced on its point is symmetrical about the vertical, but it will break that symmetry under a small perturbation by falling in a particular horizontal direction.

standard model The name for a basic theory of elementary particles, established by the mid 1970s, in which quarks and gluons (*q.v.*) are the fundamental constituents of matter and the weak and electromagnetic forces of nature are united in a single 'electroweak' theory.

synoptic gospels The first three gospels (Matthew, Mark and Luke), whose similarities of style and degree of shared material make them appear to take a common view of Jesus, in comparison with the markedly different style and material of the fourth gospel (John).

theodicy The justification of God's ways with his creation, particularly in the face of the problems of evil and suffering.

Theory of Everything (TOE) The current ambition of physical science to find a unitary theory comprehending in a single account all forms of matter and all the fundamental forces of nature.

Turing test Alan Turing proposed a behavioural test for deciding whether a computer can think: could one tell, in a conversation, conducted via a computer terminal, whether the input one was receiving originated from a person or a machine? If not, the machine passed the test.

vacuum The lowest energy state of a system. In quantum mechanics, this is still a state of activity because, for example, Heisenberg's uncertainty principle does not allow a particle to be at rest in its lowest energy position, since one would then know both its position and its momentum exactly. Consequently, there is a certain necessary and irreducible 'quiveriness' involved, which is called vacuum fluctuation.

wave and particle Quantum entities, such as electrons or photons, show both particle-like and wave-like properties in appropriate circumstances. This counter-intuitive behaviour does not lead to inconsistency, since the different behaviours are elicited by different and mutually exclusive forms of experimental investigation (see complementarity, above).

Bibliography

Allen, D. (1989), *Christian Belief in the Modern World*, Westminster/John Knox Press
Arbib, M. and Hesse, M. (1986), *The Construction of Reality*, Cambridge University Press
Baillie, D. M. (1956), *God Was in Christ*, Faber & Faber
Baker, J. A. (1970), *The Foolishness of God*, Darton, Longman & Todd
Banner, M. C. (1990), *The Justification of Science and the Rationality of Religious Belief*, Oxford University Press
Barbour, I. G. (1956), *Issues in Science and Religion*, SCM Press
—— (1974), *Myths, Models and Paradigms*, SCM Press
—— (1990), *Religion in an Age of Science*, SCM Press
Barnes, M. (1989), *Religions in Conversation*, SPCK
Barr, J. (1980), *Explorations in Theology* 7, SCM Press
Barrett, C. K. (1956), *The New Testament Background: Selected Documents*, SPCK
Barrow, J. D. (1992), *Pi in the Sky*, Oxford University Press
Barrow, J. D. and Tipler, F. J. (1986), *The Anthropic Cosmological Principle*, Oxford University Press
Bartholomew, D. J. (1984), *God of Chance*, SCM Press
Begbie, J. (1991), *Voicing Creation's Praise*, T. & T. Clark
Berger, P. (1970), *A Rumour of Angels*, Penguin
Bohm, D. (1980), *Wholeness and the Implicate Order*, Routledge & Kegan Paul
Bradley, I. (1990), *God is Green*, Darton, Longman & Todd
Bradshaw, P. (1992), *The Search for the Origins of Christian Worship*, SPCK
Bronowski, J. (1973), *The Ascent of Man*, BBC Publications
Brown, D. (1985), *The Divine Trinity*, Duckworth
—— (1987), *Continental Philosophy and Modern Theology*, Blackwell
Buckley, M. J. (1987), *At the Origins of Modern Atheism*, Yale University Press
Burrell, D. B. (1986), *Knowing the Unknowable God*, University of Notre Dame Press
Burrows, R. (1987), *Ascent to Love*, Darton, Longman & Todd
Caird, G. R. (1963), *The Gospel of Luke*, Penguin
Capra, F. (1975), *The Tao of Physics*, Wildwood House
Carnes, J. R. (1982), *Axiomatics and Dogmatics*, Christian Journals
Cobb, J. B. and Griffin, D. R. (1976), *Process Theology*, Westminster Press
Cracknell, K. (1986), *Towards a New Relationship*, Epworth Press
Cragg, K. (1986), *The Christ and the Faiths*, SPCK
Crick, F. (1966), *Of Molecules and Men*, University of Washington Press
—— (1988), *What Mad Pursuit*, Weidenfeld & Nicolson

Cullmann, O. (1963), *The Christology of the New Testament*, SCM Press
Cupitt, D. (1980), *Taking Leave of God*, SCM Press
Daly, G. (1988), *Creation and Redemption*, Gill & Macmillan
Davies, P. C. W. (1983), *God and the New Physics*, Dent
—— (1984), *Superforce*, Heinemann
—— (1987), *The Cosmic Blueprint*, Heinemann
—— (ed.) (1989), *The New Physics*, Cambridge University Press
—— (1992), *The Mind of God*, Simon & Schuster
Dawkins, R. (1986), *The Blind Watchmaker*, Longman
D'Costa, G. (1986), *Theology and Religious Pluralism*, Blackwell
—— (ed.) (1990), *Christian Uniqueness Reconsidered*, Orbis Books
Desmond, A. and Moore, J. (1991), *Darwin*, Michael Joseph
d'Espagnat, B. (1989), *Reality and the Physicist*, Cambridge University Press
Doctrine Commission (1987), *We Believe in God*, Church House Publishing
Dodd, C. H. (1961), *The Parables of the Kingdom*, James Nisbet
—— (1963), *Historical Tradition in the Fourth Gospel*, Cambridge University Press
—— (1971), *The Founder of Christianity*, Collins
Donovan, V. J. (1982), *Christianity Rediscovered*, SCM Press
Drees, W. B. (1990), *Beyond the Big Bang*, Open Court
Dunn, J. D. G. (1977), *Unity and Diversity in the New Testament*, SCM Press
—— (1980), *Christology in the Making*, SCM Press
—— (1985), *The Evidence for Jesus*, SCM Press
Dyson, F. J. (1979), *Disturbing the Universe*, Harper & Row
—— (1988), *Infinite in All Directions*, Harper & Row
Eccles, J. (1984), *The Human Mystery*, Routledge & Kegan Paul
—— (1989), *Evolution of the Brain: Creation of the Self*, Routledge
Eccles, J. and Robinson, D. N. (1985), *The Wonder of Being Human*, New Science Library
Einstein, A. (1948), *Ideas and Opinions*, Crown
Farrer, A. M. (1962), *Love Almighty and Ills Unlimited*, Collins
—— (1967), *Faith and Speculation*, A. & C. Black
—— (1968), *A Science of God?*, Geoffrey Bles
Fiddes, P. (1988), *The Creative Suffering of God*, Oxford University Press
Ford, D. F. (ed.) (1989), *The Modern Theologians*, vol. 1, Blackwell
Gilkey, L. (1959), *Maker of Heaven and Earth*, Doubleday
—— (1969), *Naming the Whirlwind*, Bobbs-Merrill
Gleick, J. (1988), *Chaos*, Heinemann
Gorringe, T. M. (1990), *Discerning Spirit*, SCM Press
—— (1991), *God's Theatre*, SCM Press
Green, C. (ed.) (1989), *Karl Barth*, Collins
Grenz, S. J. (1990), *Reason for Hope*, Oxford University Press
Griffin, D. R. (ed.) (1986), *Physics and the Ultimate Significance of Time*, State University of New York Press
Gunton, C. E. (1991), *The Promise of Trinitarian Theology*, T. & T. Clark
Gutiérrez, G. (1988), *A Theology of Liberation*, revd edn, SCM Press
Hanson, N. R. (1969), *Perception and Discovery*, Freeman Cooper
Hartshorne, C. (1948), *The Divine Relativity*, Yale University Press
Harvey, A. E. (1982), *Jesus and the Constraints of History*, Duckworth

BIBLIOGRAPHY

Hawking, S. W. (1988), *A Brief History of Time*, Bantam
Hebblethwaite, B. (1987), *The Incarnation*, Cambridge University Press
Hengel, M. (1981), *Atonement*, SCM Press
Hick, J. (1966a), *Faith and Knowledge*, Cornell University Press
—— (1966b), *Evil and the God of Love*, Macmillan
—— (ed.) (1976), *The Myth of God Incarnate*, SCM Press
—— (1989), *An Interpretation of Religion*, Macmillan
Hodgson, D. (1991), *The Mind Matters*, Oxford University Press
Jaki, S. (1978), *The Road of Science and the Ways to God*, Scottish Academic Press
—— (1989a), *Miracles and Physics*, Christendom Press
—— (1989b), *God and the Cosmologists*, Scottish Academic Press
Jantzen, G. M. (1984), *God's World, God's Body*, Darton, Longman & Todd
—— (1987), *Julian of Norwich*, SPCK
Jenkins, D. (1967), *The Glory of Man*, SCM Press
Jeremias, J. (1971), *New Testament Theology*, vol. 1, SCM Press
—— (1972), *The Parables of Jesus*, SCM Press
Jung, C. G. (1954), *Answer to Job*, Routledge & Kegan Paul
Kenny, A. (1979), *The God of the Philosophers*, Oxford University Press
—— (1989), *The Metaphysics of Mind*, Oxford University Press
Knox, J. (1967), *The Humanity and Divinity of Christ*, Cambridge University Press
Kragh, H. J. (1990), *Dirac*, Cambridge University Press
Küng, H. (1977), *On Being a Christian*, Collins
—— (1986), *Christianity and the World Religions*, Doubleday
Lakatos, I. (1978), *The Methodology of Scientific Research Programmes*, Cambridge University Press
Lampe, G. W. H. (1977), *God as Spirit*, Oxford University Press
Lapide, P. (1984), *The Resurrection of Jesus*, SPCK
Leslie, J. (1989), *Universes*, Routledge
Lewis, C. S. (1946), *The Great Divorce*, Geoffrey Bles
Lindbeck, G. A. (1984), *The Nature of Doctrine*, SPCK
Lockwood, M. (1989), *Mind, Brain and the Quantum*, Blackwell
Lonergan, B. (1972), *Method in Theology*, Darton, Longman & Todd
Lossky, V. (1957), *The Mystical Theology of the Eastern Church*, James Clarke
McDonald, J. I. H. (1989), *The Resurrection*, SPCK
MacIntyre, A. (1981), *After Virtue*, Duckworth
Mackay, A. L. (1977), *The Harvest of a Quiet Eye*, Institute of Physics
Mackey, J. P. (1987), *Modern Theology*, Oxford University Press
McMullin, E. (ed.) (1985), *Evolution and Creation*, University of Notre Dame Press
—— (ed.) (1988), *Construction and Constraint*, University of Notre Dame Press
Macquarrie, J. (1977), *Principles of Christian Theology*, SCM Press
—— (1981), *Twentieth-Century Religious Thought*, SCM Press
—— (1982), *In Search of Humanity*, SCM Press
—— (1990), *Jesus Christ in Modern Thought*, SCM Press
Mickens, R. E. (ed.) (1990), *Mathematics and Science*, World Scientific
Mitchell, B. (1973), *The Justification of Religious Belief*, Macmillan
Moltmann, J. (1967), *Theology of Hope*, SCM Press
—— (1974), *The Crucified God*, SCM Press
—— (1977), *The Church in the Power of the Spirit*, SCM Press

BIBLIOGRAPHY

—— (1981), *The Trinity and the Kingdom of God*, SCM Press
—— (1985), *God in Creation*, SCM Press
—— (1990), *The Way of Jesus Christ*, SCM Press
—— (1992), *The Spirit of Life*, SCM Press
Monod, J. (1972), *Chance and Necessity*, Collins
Montefiore, H. (1985), *The Probability of God*, SCM Press
—— (ed.) (1992), *The Gospel and Contemporary Culture*, Mowbray
Moore, W. (1989), *Schrödinger*, Cambridge University Press
Morris, T. V. (1986), *The Logic of God Incarnate*, Cornell University Press
Moule, C. D. F. (1977), *The Origin of Christology*, Cambridge University Press
Murphy, N. (1990), *Theology in the Age of Scientific Reasoning*, Cornell University Press
Nagel, T. (1986), *The View from Nowhere*, Oxford University Press
Neil, S. C. (1964), *The Interpretation of the New Testament, 1861-1961*, Oxford University Press
Newton-Smith, W. H. (1981), *The Rationality of Science*, Routledge & Kegan Paul
Niebuhr, H. Richard (1951), *Christ and Culture*, Harper & Row
O'Collins, G. (1987), *The Easter Jesus*, Darton, Longman & Todd
Pailin, D. A. (1989), *God and the Process of Reality*, Routledge
—— (1992), *A Gentle Touch*, SPCK
Pannenberg, W. (1968), *Jesus – God and Man*, SCM Press
—— (1976), *Theology and the Philosophy of Science*, Darton, Longman & Todd
Park, D. (1988), *The Why and the How*, Princeton University Press
Pattison, G. (1991), *Art, Modernity and Faith*, Macmillan
Peacocke, A. R. (1979), *Creation and the World of Science*, Oxford University Press
—— (1984), *Intimations of Reality*, University of Notre Dame Press
—— (1986), *God and the New Biology*, Dent
—— (1990), *Theology for a Scientific Age*, Blackwell
Penrose, R. (1989), *The Emperor's New Mind*, Oxford University Press
Peters, T. (ed.) (1989), *Cosmos and Creation*, Abingdon Press
Phillips, D. Z. (1976), *Religion Without Explanation*, Blackwell
Polanyi, M. (1958), *Personal Knowledge*, Routledge & Kegan Paul
—— (1969), *Knowing and Being*, Routledge & Kegan Paul
Polkinghorne, J. C. (1979), *The Particle Play*, W. H. Freeman
—— (1983), *The Way the World Is*, Triangle/Eerdmans
—— (1984), *The Quantum World*, Longman/Princeton University Press
—— (1986), *One World*, SPCK/Princeton University Press
—— (1988), *Science and Creation*, SPCK/New Science Library
—— (1989a), *Science and Providence*, SPCK/New Science Library
—— (1989b), *Rochester Roundabout*, Longman/W. H.Freeman
—— (1991), *Reason and Reality*, SPCK/Trinity Press International
Pollard, W. G. (1958), *Chance and Providence*, Faber & Faber
Popper, K. (1959), *The Logic of Scientific Discovery*, Hutchinson
—— (1963), *Conjectures and Refutations*, Routledge & Kegan Paul
Prigogine, I. (1980), *From Being to Becoming*, W. H. Freeman
Prigogine, I. and Stengers, I. (1984), *Order Out of Chaos*, Heinemann
Puddefoot, J. (1987), *Logic and Affirmation*, Scottish Academic Press
Race, A. (1985), *Christians and Religious Pluralism*, SCM Press

Robinson, J. A. T. (1950), *In the End God*, James Clarke
——— (1952), *The Body*, SCM Press
——— (1972), *The Human Face of God*, SCM Press
——— (1976), *Redating the New Testament*, SCM Press
——— (1979), *Truth is Two-Eyed*, SCM Press
——— (1984), *Twelve More New Testament Studies*, SCM Press
——— (1985), *The Priority of John*, SCM Press
Rolston, H. (1987), *Science and Religion*, Temple University Press
Rorty, R. (1980), *Philosophy and the Mirror of Nature*, Blackwell
Rowland, C. and Corner, M. (1990), *Liberating Exegesis*, SPCK
Russell, C. A. (1986), *Cross-Currents*, IVP
Russell, R. J., Stoeger, W. R. and Coyne, C. V. (ed.) (1988), *Physics, Philosophy and Theology*, Vatican Observatory
Russell, R. J., Murphy, N. and Isham, C. J. (ed.) (1993), *Quantum Cosmology and the Laws of Nature*, Vatican Observatory
Sagan, C. (1980), *Cosmos*, Random House
Sanders, E. P. (1977), *Paul and Palestinian Judaism*, SCM Press
——— (1985), *Jesus and Judaism*, SCM Press
Sanders, E. P. and Davies, M. (1989), *Studying the Synoptic Gospels*, SCM Press
Schillebeeckx, E. (1974), *Jesus*, Collins
Searle, J. (1984), *Minds, Brains and Science*, BBC Publications
Sherwin-White, A. N. (1963), *Roman Society and Roman Law in the New Testament*, Oxford University Press
Stannard, R. (1982), *Science and the Renewal of Belief*, SCM Press
Steiner, G. (1989), *Real Presences*, Faber & Faber
Stewart, I. (1989), *Does God Play Dice?*, Blackwell
Swinburne, R. G. (1979), *The Existence of God*, Oxford University Press
Sykes, S. W. (1984), *The Identity of Christianity*, SPCK
Taylor, J. V. (1972), *The Go-Between God*, SCM Press
——— (1992), *The Christlike God*, SCM Press
Taylor, W. S. (n.d.), *The Far Side of Reason*, G. R. Welch
Teilhard de Chardin, P. (1959), *The Phenomenon of Man*, Collins
Temple, W. (1924), *Christus Veritas*, Macmillan
Theissen, G. (1984), *Biblical Faith*, SCM Press
Thiemann, R. E. (1985), *Revelation and Theology*, University of Notre Dame Press
Thiselton, A. C. (1980), *The Two Horizons*, Paternoster Press
Thomson, A. (1987), *Tradition and Authority in Science and Theology*, Scottish Academic Press
Torrance, T. F. (1969), *Theological Science*, Oxford University Press
——— (1980), *The Ground and Grammar of Theology*, University Press of Virginia
——— (1981), *Divine and Contingent Order*, Oxford University Press
——— (1985), *Reality and Scientific Theology*, Scottish Academic Press
——— (1989), *The Christian Frame of Mind*, Helmers & Howard
Tracy, D. (1981), *The Analogical Imagination*, SCM Press
van Fraassen, B. (1980), *The Scientific Image*, Oxford University Press
——— (1989), *Laws and Symmetry*, Oxford University Press
Vanstone, W. H. (1977), *Love's Endeavour, Love's Expense*, Darton, Longman & Todd
——— (1982), *The Stature of Waiting*, Darton, Longman & Todd

BIBLIOGRAPHY

Vermes, G. (1983), *Jesus the Jew*, SCM Press
Vidler, A. (ed.) (1962), *Soundings*, Cambridge University Press
Ward, K. (1982a), *Rational Theology and the Creativity of God*, Blackwell
—— (1982b), *Holding Fast to God*, SPCK
—— (1987), *Images of Eternity*, Darton, Longman & Todd
—— (1990), *Divine Action*, Collins
—— (1991), *A Vision to Pursue*, SCM Press
Weber, R. (1980), *Dialogues with Sages and Scientists*, Routledge & Kegan Paul
Weinberg, S. (1977), *The First Three Minutes*, A. Deutsch
—— (1993), *Dreams of a Final Theory*, Hutchinson Radius
White, V. (1985), *The Fall of a Sparrow*, Paternoster Press
—— (1991), *Atonement and Incarnation*, Cambridge University Press
Whitehead, A. N. (1978), *Process and Reality*, The Free Press
Wiles, M. F. (1986), *God's Action in the World*, SCM Press
—— (1992), *Christian Theology and Inter-religious Dialogue*, SCM Press
Williams, R. (1991), *Teresa of Avila*, Geoffrey Chapman
Wolpert, L. (1992), *The Unnatural Nature of Science*, Faber & Faber.
Wright, N. T. (1992), *The New Testament and the People of God*, SPCK
Young, F. M. (1975), *Sacrifice and the Death of Christ*, SPCK
Zizioulas, J. (1985), *Being as Communion*, Darton, Longman & Todd
Zukav, G. (1979), *The Dancing Wu-Li Masters*, Hutchinson

Index

INDEX